Center for Basque Studies
Current Research Series, No. 10

Sustainable Development, Ecological Complexity, and Environmental Values

Edited by

Ignacio Ayestarán and Miren Onaindia

Current Research Series No. 10

Center for Basque Studies
University of Nevada, Reno

Published in conjunction with the University of the Basque Country
UPV/EHU

Center for Basque Studies
UNIVERSITY OF NEVADA, RENO

eman ta zabal zazu
Universidad del País Vasco
Euskal Herriko Unibertsitatea

Current Research Series No. 10
Center for Basque Studies
University of Nevada, Reno
Reno, Nevada 89557
http://basque.unr.edu

Cover design based on engravings by Eduardo Chillida and Jorge Oteiza.
Translations by: introduction and chapter 2, Jennifer Martin; chapters 1, 10, and 13 Julie Waddington; chapter 3 and 4, Lauren DeAre; chapters 6 and 12, Laura Bunt-MacRury; chapters 7, 9, and 11, Jennifer Ottman

Library of Congress Cataloging-in-Publication Data

Sustainable development, ecological complexity, and environmental values / edited by Ignacio Ayestarán and Miren Onaindia.
pages cm. -- (Current research series ; no. 10)
"Published in conjunction with the University of the Basque Country."
Summary: "Collection of articles on sustainability in the Basque Country"--Provided by publisher.
ISBN 978-1-935709-35-0 (pbk.)
1. País Vasco (Spain)--Environmental conditions. 2. Pays Basque (France)--Environmental conditions. 3. Sustainability--Spain--País Vasco. 4. Sustainability--France--Pays Basque. 5. Sustainable development--Spain--País Vasco. 6. Sustainable development--France--Pays Basque. I. Ayestarán, Ignacio. II. Onaindia, Miren. III. University of Nevada, Reno. Center for Basque Studies. IV. Universidad del País Vasco.

GE160.S7S87 2013
338.946'607--dc23

2013005092

Contents

Introduction

IGNACIO AYESTARÁN and MIREN ONAINDIA
The University of the Basque Country (UPV/EHU)

In 1987, the World Commission on Environment and Development (better known as the Brundtland Commission) published its famous report *Our Common Future*, in which, for the first time, the concept of *sustainable development* was internationally defined as development that meets the needs of the present without compromising the ability of future generations to meet their own needs. In 2007, the United Nations Environment Programme (UNEP) published the assessment report *Global Environment Outlook: Environment for Development* (GEO-4), to determine how our planet had changed after two decades of concern over sustainable development. According to the UNEP, the world has radically changed on a social, economic, and environmental scale since 1987: the world population grew from 5 billion to 7 billion people, trade is nearly three times greater, and the average per capita income increased by almost 40 percent. But it did so unequally and with broad margins of environmental degradation of ecosystems within several indicators (water management; loss of biodiversity; urban development and population; climate change of anthropogenic origin; agrarian, livestock, and fishing overexploitation; among others). The degradation of ecosystem services may considerably worsen on the planet during the first half of this century, according to the United Nations' reports on *Millennium Ecosystem Assessment*. This would be a major obstacle in achieving the Millennium Development Goals of reducing poverty, hunger, and diseases, for which environmental sustainability is fundamental.

On the global scale, changes have clearly accelerated across the entire planet. From a global point of view, ecological and social sustainability are two facets of the same changing reality. On the one hand, social sustainability depends on ecological sustainability: If we continue degrading nature's capacity to produce the ecosystems' services (water filtration, climate stabilization, etc.) and resources (food, materials), both individuals and nations will be affected by growing pressures and increasing conflicts, as well as by threats to public health and personal safety. On the other hand, ecological sustainability depends on social sustainability: With a growing population living in a social system that does not make the fulfillment of its needs possible, it is increasingly difficult to protect the natural environment. The forests are felled for agriculture, the pastures are overexploited, the aquifers degraded, the rivers and seas overused, although some of nature may be conserved in small reserve areas or natural parks. In addition, human behavior and the social dynamic often lie at the heart of social and ecological problems. It must be, therefore, assumed that there will not be sustainable development if sustainable societies do not first exist.

A sustainable society has the challenge of developing human capital in the areas of education, health, job creation, and innovation. This process has ethical implications both in the areas of transgenerational solidarity (with future generations) and intragenerational solidarity (with current generations from the most disadvantaged places), as well as on issues related to gender equality and opportunities in a complex and global world.

It is within this context that the present text is meant to be a contribution toward sustainability, integrating different thematic issues related to sustainable development in its threefold consideration (economic, social, and environmental). Because reality always appears to us as a complex and transdisciplinary connection, the present effort is part of a multidisciplinary approach that includes methodology, knowledge and the complexity of values.

1

An Evaluation of Ecosystem Services as a Base for the Sustainable Management of a Region

Miren Onaindia and Gloria Rodríguez-Loinaz

The University of the Basque Country (UPV/EHU)

Ecosystems constitute a natural capital that needs to be conserved to ensure that different services are available, such as climate control, carbon fixing, land fertility, pollination, filtration of pollutants, clean water provisions, flood controls, recreation, and aesthetic and spiritual values (Daily 1997). These ecosystem services have direct consequences on the prosperity of human society not only in terms of its economy, but also in relation to health, social relations, freedom, and safety (Millennium Ecosystem Assessment 2005, v).

Many ecosystem services are considered to be free and unlimited. However, the noncommercialized benefits are generally higher and sometimes of greater worth than the commercial ones. When ecosystem services are taken into account, the value of natural and sustainably managed ecosystems is frequently greater than that of converted or intensively managed systems. For example, a comprehensive study that examined the marketed and nonmarketed economic values associated with forests in eight Mediterranean countries found that timber and fuelwood generally accounted for less than a third of the total economic value of the forests in each country. Values associated with nonwood forest products (recreation, watershed protection, carbon sequestration, etc.), accounted for between 25 percent and 96 percent of the total economic value of the forests (Millennium Ecosystem Assessment 2005, 6). In different places across the world, natural ecosystems provide societies with great benefits, as is the case with the mangrove swamps (Barbier 2007).

* We would like to thank the following bodies for the funding provided to carry out this research: the Ministry of Science and Innovation, project CGL2008-05579-C02-01/BOS and the Departament of Education, Universities and Research of the Basque Government (funding for research groups, 2009 funding round).

The degradation of ecosystem services represents a loss of natural capital, even though this is not reflected in the conventional indices for measuring income. For example, a country may chop down all of its forests and put an end to fishing and still increase its gross domestic product (Perrings 2005).

Conservation policies and those related to the management of natural resources have generally been based on strategies aimed at planning in a specific sector without consideration of the potential overall effects on the environment and society. For example, maximizing agricultural production may lead to the degradation of water quality in rivers and aquifers (Tallis and Polasky 2009, 265).

The sustainable management of a region requires the integration of ecological, social, economic, and institutional perspectives, based on the recognition of the important interdependence that exists between them (Pikitch et al. 2004, 346). The paradigm underpinning any ecosystem approach to management policies requires a total interrelation between human and ecological well-being in such a way that sustainability is only possible if both aspects are considered simultaneously. This theoretical approach enables managers to obtain a wider perspective of the multiple consequences that can result from specific decisions (Christensen et al. 1996, 681).

Over recent years, large-scale changes to ecosystems, such as the conversion of natural systems into single-crop agricultural systems, has led to an increase in the provision of certain services (e.g., food production), at the expense of different ecosystem regulatory services and cultural services (Vitousek et al. 1997). As a result, awareness of the distribution of these services is very useful and informative when making decisions about management issues. It is also necessary to share experiences in the study and application of ecosystem services and to define priorities for future work (Daily and Matson 2008, 9455), since the relevant areas for the provision of ecosystem services should be managed in a sustainable way to ensure the present provisions as well as guarantee the future provision of these services (Egoh et al. 2007, 718–19).

Based on the conceptual framework proposed, the objective of this study is to evaluate the distribution of specific ecosystem services as a basis for the sustainable management of a region. To do this, the ecosystem services and their distribution across the region are evaluated by identifying the areas with the greatest value in maintaining specific natural and cultural values of the ecosystems. Thus, four ecosystem services have been selected: biodiversity (support services); regulation of the hydrological cycle; storage of carbon both in soil and vegetation (both regulatory services); and recreational use (cultural services) (de Groot, Wilson, and Boumans 2002, 396–97).

The relationship between the distribution of services—especially between biodiversity and the other services studied—is analyzed to determine whether it could be a target parameter (Sarkar et al. 2005, 816) for the evaluation of the other services. On the one hand, the distribution of the vegetation growth rate has been analyzed as a measure of production to test its capacity as a means of evaluating the synthesis of ecosystem services.

Methodology

Area of Study

The area of this study is centered on the Urdaibai Biosphere Reserve in the province of Bizkaia (Vizcaya), and follows a particular kind of sustainable development. This is an area of great value, both in terms of its landscapes and its ecological diversity, which is in a fairly urbanized area not far from the metropolitan region of Bilbao (with more than 1 million inhabitants). Urdaibai covers an area of 22,041 hectares, including twenty two municipalities, and comprises approximately 10 percent of the surface of Bizkaia, and 3 percent of the Comunidad Autónoma del País Vasco-Euskal Autonomia Erkidegoa (CAPV-EAE, Autonomous Community of the Basque Country). The most important municipalities are Gernika (Guernica) inland and Bermeo on the coast (figure 1.1). A great diversity of ecosystems and habitats can be found in this region (dunes, Cantabrian holm oak groves, marshlands, Atlantic countryside, forestry, agricultural crops, etc.), which encourages the existence of a high level of biodiversity, particularly in relation to birds (IKT and the Governing Committee of the Urdaibai Reserve, 2006). On the other hand, the population of the municipality is close to forty-five thousand inhabitants, whose economic activity is based on the metallurgical and primary sectors (fishing, agriculture, livestock, and forestry). Moreover, the landscape of the reserve is a highly valued resource. Its location generates significant conflicts of interest in terms of land use, making it an area of particular interest for the application of a natural resource management strategy based on ecosystem services.

Mapping and Evaluating Ecosystem Services

The cartography of the ecosystem services is based on an analysis of habitat, inclinations, altitude, and lithology maps. The database used for the analysis has been obtained from orthophotos with a scale of 1:10,000 that were provided by the cartography service of the Provincial Council of Bizkaia. The information has been completed with field sampling according to the habitat classes defined in the EUNIS Habitat classification (Rodríguez et al. 2007). Areas altered by human activity (cities, roads, parks) and water masses are not taken into account in this particular study.

Figure 1.1. Location of the study area.

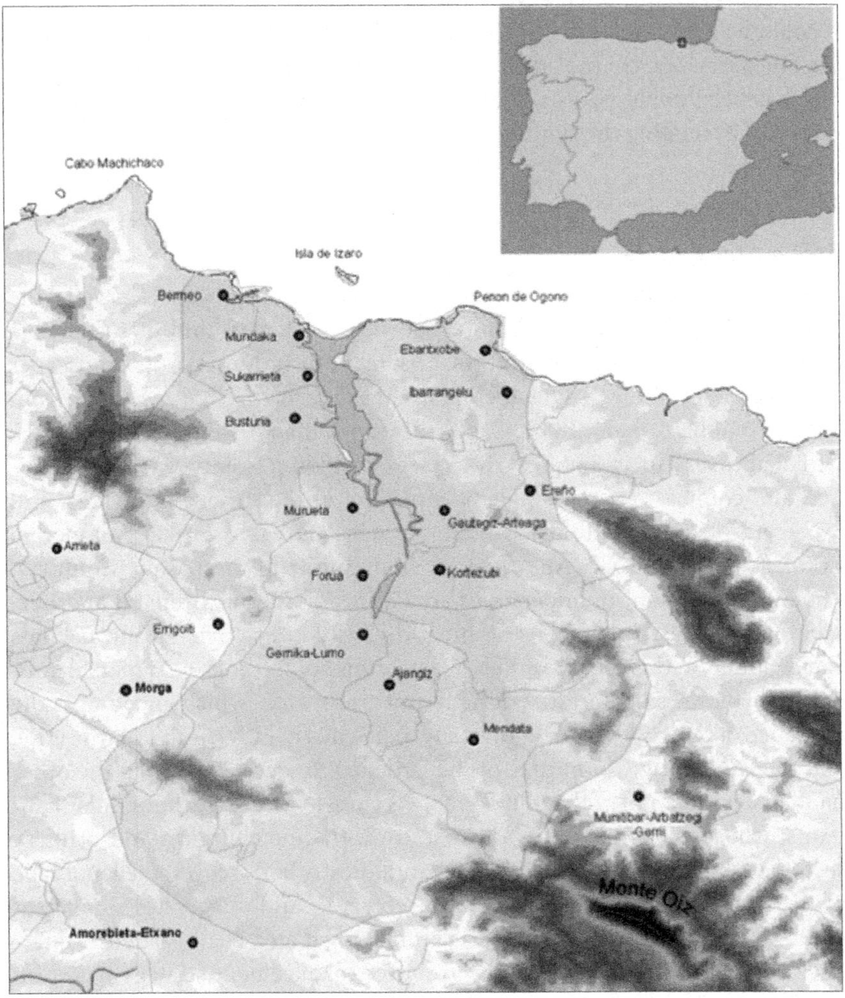

The assessment of the different aspects covered is carried out on a relative scale of four levels, ranging from one to four (One representing a very low value and 4 representing the highest attainable score within the ranges of this study). The highest value of a specific service is considered a "hotspot," indicating an area whose management is of utmost importance for the service in question (Egoh et al. 2008, 136). The evaluation of biodiversity is carried out by calculating wealth—which refers to the total number of vegetable species found on the land (vascular plants)—by way of the field sampling of the environmental units specified (Rodríguez et al. 2007).

The capacity to regulate hydrological flow is directly proportional to the volume of water retained or stored in the soil and vegetation. To a large extent, the regulation of the hydrological cycle depends on plant coverage, the permeability of the geological substratum, the state of the soil, the state of the river basin (the area at the top or bottom of the valley), and the incline. Therefore, in addition to vegetation maps, for this evaluation lithology, altitude, and incline maps were also used.

For the calculation of carbon storage, the data obtained in a report carried out by the Basque Institute for Agricultural Research and Development, Neiker (2004) has been used. In this section, the carbon stock in biomass and in soil has been assessed separately. For the calculation of the growth rate, the rates corresponding to the dominant species in each environmental unit have been used, and these are expressed in Gr $ha^{-1}year^{-1}$. The data has been obtained from the IPCC (2004).

The evaluation of the importance of environmental units as recreational areas is carried out by consulting different groups who are asked to score the environmental units on a scale of 1 to 4 according to their preferences. Fifty surveys have been carried out for this aspect. The final score for each unit was obtained by calculating the average value of the scores of the different groups for said unit (Casado, Palacios, and Onaindia 2010).

Once the services have been evaluated, the distribution of these services is represented cartographically in the space provided and any overlaps in the distribution of the services studied are calculated (Egoh et al. 2008, 137) using a referenced geographical information system (GIS).

Results

An Evaluation of the Ecosystem Services

The greatest wealth is present in natural hardwood forests, whether deciduous or evergreen, as well as gallery forests. The wealth of species is gradually diminishing in plantations as a result of the management they are subjected to, particularly in the case of conifer plantations, thus preventing the development of nemoral species. As a result, natural and gallery forests score 4, mature hardwood forests score 3, and conifer plantations score just 2. Eucalyptus plantations, which are planted in dense populations, receive a score of 1 because they exclude all other species. A score of 1 has been given to all young forest plantations, whether hardwood or conifer, due to the low number of species found in them. Meadows score 4, as they present a great wealth of plant diversity (table 1.1).

In terms of the regulation of the hydrological cycle, areas with a highly permeable substratum (e.g., karst outcrops), act as huge sponges for the recharge of aquifers, and therefore score particularly well. Natural forest

Table 1.1. Environmental units defined and their scores for the different ecosystem services studied. Synthetic areas and water masses are not taken into account in this assessment.

Assessment				
	Support systems	Regulatory systems	Cultural systems	
	Biodiversity	Carbon stock (biomass/ soil)	Recreational	Growth rate
Natural vegetation of marshlands	2	3	2	1
Invasive vegetation of marshlands	1	2 (0/2)	1	2
Coastal sandbanks	2	1	2	1
Coastal cliffs	2	1	2	1
Reedbeds	1	2	2	1
Meadows	4	2 (0/2)	3	1
Fern vegetation	2	2 (0/2)	2	1
Bushes	3	3 (1/2)	2	1
Riverside forests	4	4 (2/2)	3	4
Cantabrian holm oak groves	4	3 (1/2)	4	1
Mature deciduous forests	4	4 (2/2)	4	1
Young deciduous forests	4	2 (1/1)	3	3
Mature hardwood plantations	3	4 (2/2)	3	2
Young hardwood plantations	1	2 (1/1)	2	3
Eucalyptus plantations	1	3 (1/2)	1	4
Mature conifer plantations	2	4 (2/2)	2	2
Young conifer plantations	1	2 (1/1)	1	4

formations in zones at the head of and on elevated inclines also score well because they play an important role in protecting the river basin. Forests in general are highly valued because they reduce water runoff and encourage infiltration. Nevertheless, intensive forest plantations score lower because they are normally associated with the use of machinery that leads to soil compaction. Finally, the lowest score is given to areas at the bottom of valleys with little incline, given that there is no problem of surface runoff in these areas, meaning that this service is not created (table 1.2).

With regard to carbon storage, forest ecosystems play a vital role in the carbon cycle, constituting one of the largest reserves and sinks of carbon. As a general rule, species that grow quickly, such as pine and eucalyptus, tend to fix carbon more quickly, which also occurs with young specimens as opposed to old ones. When they grow, plant species extract carbon from the atmosphere through photosynthesis and accumulate it in their tissues, thereby fix-

Table 1.2. Assessment of the regulatory service of the hydrological cycle of the different areas.

		Assessment Regulation of hydrological cycle
Karst outcrop areas (regardless of the kind of plant coverage)		4
Headwater areas (above 150 m)	Semi-natural forests and hardwood plantations	3
	Conifer and eucalyptus plantations and bushes	2
	Meadows and fern vegetation	1
Areas at the bottom of valleys (under 150 m) (regardless of the plant coverage)		1

ing it. The quantity of carbon fixed in mature forests is greater than that in young ones (table 1.1).

In terms of the assessment of recreational use, the results show that society values highly mature hardwood forests, with natural forests receiving a score of 4, and plantations a score of 3. Conifer plantations are less valued, with a score of 2 (table 1.1).

Spatial Distribution of Ecosystem Services

The spatial distribution of ecosystem services in the area of the Urdaibai Reserve is extensive for all the services studied, given that at least one of the services is provided in 90 percent of the region. The highest values of biodiversity (the support services of the ecosystems) represent 34.7 percent of the total surface, with the highest values being 4.4 percent, the average value being 33 percent, and the lowest value being 21 percent. With regards to recreational use (the cultural services of the ecosystems), the areas with very high values represent 10.6 percent of the surface area, the next highest values 26.2 percent, the average 36.7 percent, and the lowest 19.6 percent. The region's highest values of carbon storage (the regulatory services of the ecosystems) represent 36.7 percent of the surface area, with the highest values being 17.7 percent, the average value being 38.3 percent, and the lowest value being 0.4 percent. The regulatory services of the hydrological cycle (the regulatory services of the ecosystems) have a very high value across 11 percent of the surface area, with high values of 8.5 percent, averages of 44.2 percent, and low scores of 29.4 percent. The growth rate scores are distributed as follows: very high scores in 19.7 percent of the region, high scores in 5.5 percent, average scores in 32.7 percent, and low scores in 35.2 percent.

The most significant overlap in the spatial distribution of the hotspots is evident between biodiversity and recreational use, which amounts to 98 percent. In other words, 98 percent of the hotspot area for recreational use is found within the highest scoring area for biodiversity. The overlap between biodiversity and the regulation of the hydrological cycle is also very high, standing at 70.3 percent, and also relatively significant in terms of carbon storage (13 percent). However, the overlap between the distribution areas of the maximum scores for growth rate are very low for all of the services studied: 0 percent overlap in relation to recreational use, 1.5 percent with biodiversity, 6 percent with the regulation of the hydrological cycle, and 1.7 percent with carbon storage (table 1.3).

Discussion

From the results obtained, it can be deduced that most parts of the surface area of the reserve (close to 90 percent) are significant in terms of generating ecosystem services. Moreover, the hotspot areas also make up a high percentage of the surface area of the biosphere reserve, standing at 35 percent of the region in the case of biodiversity and carbon storage, values that are high compared to other regions (Egoh et al. 2008).

These results should be complemented by the value that these services provide society with beneficiaries, as well as more awareness of their distribution among beneficiaries so that managers' decisions may lead to situations characterized by greater social justice (Tallis and Polasky 2009, 266–67). It is important to consider how people, who are frequently excluded from analyses in regional planning, assess such aspects (Alessa, Kliskey, and Brown 2008; Raymond et al. 2009, 1302). In various cases, the Millennium Ecosystem Assessment study highlighted the conflict of interests among users who used the provision of ecosystem services (food, wood, etc.) as opposed to the conservation of other regulatory and cultural services. A framework of analysis that evaluates biodiversity alongside other ecosystem services could help to identify strategies and situations that would result in a win-win outcome (Tallis and Polasky 2009, 271).

The relationship between biodiversity and the distribution of ecosystem services varies depending on the characteristics of the area. As a result, the whole range of services cannot be planned solely on the basis of the distribution of just one of them (Chan et al. 2006). However, there is a high overlap in the region studied between regions of greatest biodiversity and areas with a high value of ecosystem services: the regulation of the hydrological cycle, cultural services, and carbon fixing.

The high spatial correlation between the services studied shows that biodiversity can be applied in this zone as a target parameter of ecosystem services. This has significant implications for the management of the region because it suggests a need to conserve the areas with the greatest biodiversity to preserve hydrological control, the accumulation of carbon in biomass and soil, and the recreational use of the region. These results could lead to a restriction of certain activities in areas identified as priority areas for ecosystem services, as well as to the establishment of measures for ecological mitigation and regeneration. By contrast, with regard to the growth rate as a measure of primary productivity, this cannot be used as an indicator of ecosystem services in general as its distribution is different from the other services studied.

Ecosystem services contribute to quality of life in numerous ways, both directly and indirectly. Therefore, as argued in various international forums, methodologies based on approaches that encourage a holistic understanding of the values of the region need to be applied to put sustainable planning into practice in the region (Kumar 2010).

References

Alessa, Lilian, Andrew Kliskey, and Gregory Brown. 2008. "Social-ecological Hotspots Mapping: A Spatial Approach for Identifying Coupled Social-ecological Space." *Landscape and Urban Planning* 85, no. 1: 27–39.

Barbier, Edward B. 2007. "Valuing Ecosystem Services as Productive Inputs." *Economic Policy* 22, no. 49: 177–229.

Casado, Izaskun, Igone Palacios, and Miren Onaindia. 2010. "El Cinturón Verde de Bilbao Metropolitano." *Sustrai* 91: 68–73.

Chan, Kai M.A., Rebecca M. Shaw, David R. Cameron, Emma C. Underwood, and Gretchen C. Daily. 2006. "Conservation Planning for Ecosystem Services." *PLoS Biology* 4, no. 11: 2138–52.

Christensen, Norman L., Ann M. Bartuska, James H. Brown, Stephen Carpenter, Carla D'Antonio, Robert Francis, Jerry F. *Franklin, James A. MacMahon, Reed F. Noss, David J. Parsons, Charles H. Peterson, Monica G. Turner,* and Robert G. *Woodmansee.* 1996. "The Report of the Ecological Society of America Committee on the Scientific Basis for Ecosystem Management." *Ecological Applications* 6, no. 3: 665–91.

Daily, Gretchen C. 1997. *Nature's Services: Societal Dependence on Natural Ecosystems.* Washington, D.C.: Island Press.

Daily, Gretchen C., and Pamela A. Matson. 2008. "Ecosystem Services: From Theory to Implementation." *Proceedings of the National Academy of Sciences of the United States of America* 105, no. 28: 9455–56.

de Groot, Rudolf S., Matthew A. Wilson, and Roelof M.J. Boumans. 2002. "A Typology for the Classification, Description and Valuation of Ecosystem Functions, Goods and Services." *Ecological Economics* 41, no. 3: 393–408.

Egoh, Benis, Mathieu Rouget, Belinda Reyers, Andrew T. Knight, Richard M. Cowling, Albert S. van Jaarsveld, and Adam Welz. 2007. "Integrating Ecosystem Services into Conservation Assessments: A Review." *Ecological Economics* 63, no. 4: 714–21.

Egoh, Benis, Belinda Reyers, Mathieu Rouget, David M. Richardson, David C. Le Maitre, and Albert S. van Jaarsveld. 2008. "Mapping Ecosystem Services for

Planning and Management."*Agriculture, Ecosystems and Environment* 127, nos. 1–2: 135–40.

IKT S.A., and Patronato de la Reserva de Urdaibai. 2006. *Ficha informativa de los humedales ramsar (FIR). Ría de Mundaka-Gernika.* At www.magrama.gob.es/es/biodiversidad/temas/conservacion-de-humedales/26_fir_mundaka_gernika_febrero2006_tcm7-19303.pdf.

IPCC. 2004. "Good Practice Guidance for Land Use, Land-use Change and Forestry." *IPCC National Greenhouse Gas Inventories Programme.* Kanagawa, Japan.

Millennium Ecosystem Assessment. 2005. *Ecosystem and Human Well-Being: Synthesis.* Washington, D.C.: Island Press.

Neiker. 2004. *Estudio sobre la potencialidad de los suelos y la biomasa de zonas agrícolas: Pascícolas y forestales de la CAPV como sumideros de carbono.* Report.

Perrings, Charles. 2005. "Economics and the Value of Ecosystem Services." Paper presented at the International Scientific Conference "Biodiversity Science and Governance," January 24–28, UNESCO, Paris.

Pikitch, Ellen K., Cristine Santora, Elizabeth A. Babcock, Andrew Bakun, Ramon Bonfil, David O. Conover, Paul Dayton, Phaedra Doukakis, David Fluharty, Burr Heneman, Edward D. Houde, J. Link, Patricia A. Livingston, Marc Mangel, Murdoch K. McAllister, John Pope, and Keith J. Sainsbury. 2004. "Ecosystem-based Fishery Management." *Science* 305, no. 5682 (July 16): 346–47.

Raymond, Christopher M., Brett A. Bryan, Darla H. MacDonald, Andrea Cast, Sarah Strathearn, Agnes Grandgirard, and Tina Kalivas. 2009. "Mapping Community Values for Natural Capital and Ecosystem Services." *Ecological Economics* 68, no. 5: 1301–15.

Rodríguez-Loinaz, Gloria, Ibone Amezaga, Mikel San Sebastián, Lorena Peña, and Miren Onaindia. 2007. "Análisis del paisaje de la reserva de biosfera de Urdaibai." *Forum de Sostenibilidad* 1: 59–69.

Sarkar, Sahotra, James Justus, Trevon Fuller, Chris Kelley, Justin Garson, and Michael Mayfield. 2005. "Effectiveness of Environmental Surrogates for the Selection of Conservation Area Networks." *Conservation Biology* 19, no. 3: 815–25.

Tallis, Heather, and Stephen Polasky. 2009. "Mapping and Valuing Ecosystem Services as an Approach for Conservation and Natural-Resource Management." *Annals of the New York Academy of Sciences* 1162: 265–83.

Kumar, Pushpam, ed. 2010. *The Economics of Ecosystems and Biodiversity: Ecological and Economic Foundations.* London: Earthscan.

Vitousek, Peter M., Harold A. Mooney, Jane Lubchenco, and Jerry M. Melillo. 1997. "Human Domination of Earth's Ecosystems." *Science* 277, no. 5325 (July 25): 494–99.

2

An Evaluation of Millennium Ecosystems from the Basque Country

IGONE PALACIOS, IZASKUN CASADO-ARZUAGA
University of the Basque Country (UPV/EHU)
IOSU MADARIAGA
University of the Basque Country (UPV/EHU) and Environmental Department of Bizkaia
XABIER ARANA
Environmental Department of Bizkaia

Human beings are closely interrelated with nature and with the environment. Nature provides for the needs of human beings through the functions and services of ecosystems through water, food, energy, and materials. In addition to these and other direct contributions, natural ecosystems provide different cultural services (recreational, educational, spiritual) and regulating services (regulating climate, floods, disease, water quality, water cycles, carbon capture) (Millennium Ecosystem Assessment 2005a, v), which are indirect contributions to human well-being. This concept of ecosystem services, originally proposed by John Holdren (Daily 1997, 3) and analyzed in depth in the Millennium Ecosystem Assessment and its successive phases (MA Follow-up), grew out of the paradigm that society should better understand and value ecosystems because human beings depend on them (Onaindia 2007, 41).

The United Nations (UN) launched the Millennium Ecosystem Assessment (MA) in 2001. Its objective was to contribute to the achievement of the Millennium Development Goals through better knowledge and valuation of ecosystems. The MA was carried out between 2001 and 2005 to assess the consequence of ecosystem changes on human well-being by bringing the best available infor-

* We wish to thank the Provincial Government of Bizkaia, particularly the Department of the Environment and all of the technical experts who participated in this project; the Basque Government Department of Education, Universities and Research that financed Izaskun Casado-Arzuaga's Research Fellowship; Pilar Barraqueta and Gurutxe Santos from the EKOS company for their work on the Huella Ecológica (Ecological Footprint) study; the team at Prospektiker company for its support of the scenario workshops; Carlos Montes, Berta Martín-López, the Social-Ecological Systems Laboratory from the Autonomous University of Madrid; UNESCO Etxea and all of the people who participated in the project whether by responding to the survey or as participants in the various workshops that were conducted.

mation and knowledge on ecosystem services to bear on policy and management decisions. Thus, the MA established the scientific basis for the action needed to enhance the conservation and sustainable use of ecosystems and for understanding how ecosystems contribute to human well-being (Ash et al. 2010).

Despite the efforts of various social sectors to promote nature conservation, especially during the 1970s (e.g., Meadows et. al. 1972), environmental deterioration and biodiversity loss are increasing. According to the MA report *Ecosystems and Human Well-being: Current State and Trends* (2005b), 60 percent of ecosystem services are being degraded or used in unsustainable ways, including freshwater resources, capture fisheries, air and water purification, local and regional climate regulation. Changes in the ecosystems are also increasing the likelihood of nonlinear and potentially sudden changes that carry significant consequences for human well-being such as natural risks and epidemics. The report concludes that global pressures on ecosystems will increase in the coming decades, unless human attitudes and actions change.

A report from the World Wildlife Fund, the Zoological Society of London, and Global Footprint Network (2008, 2) affirms that humanity's ecological footprint[1] has more than doubled since 1961, and is now exceeding the planet's capacity to regenerate in almost 30 percent of the cases. These findings prove that the messages used thus far in favor of nature conservation and sustainability have not made a sufficient impression on the vast majority of society nor, consequently, on the majority of its leaders. Thus, a new method of understanding the social–environmental problems is necessary, so that we can stop ignoring the obvious and begin to focus on and think about that which affects our lives and our well-being: ecosystems,[2] natural capital,[3] and our way of relating to each other and the environment that surrounds us.

Conceptual Framework and the Importance of the Millennium Ecosystem Assessment

The MA incorporates a systemic and noncompartmentalized view of the sustainability concept. It departs from the current, more advanced views of sustain-

1. A sustainability index has been developed to estimate humankind's impact on the planet and is defined by its authors as: "the area of ecologically productive land (and water) in various classes—cropland, forests, pasture, etc.—that would be required on a continuous basis (a) to provide all the energy/material resources consumed, and (b) to absorb all the waste discharged by that population with prevailing technology, *wherever on Earth that land is located*." (Wackernagel and Rees 1996, 51–52)

2. An *ecosystem* is a complex dynamic of plant, animal, and microorganism communities and the nonliving environment interacting as a functional unit. Humans are an integral part of ecosystems (Millennium Ecosystem Assessment 2005c, 25).

3. *Natural capital* refers to those systems that have the capacity to exert functions and therefore provide services to society (Martín-López et al. 2009, 263).

ability, which regard nature and human beings as a complex social-ecological system (Berkes, Colding, and Folke 2003). That is, the human system forms part of the biophysical system and, therefore, the planet's biophysical limits cannot be ignored because society and its economy depend on the functions and services of their natural ecosystems. From this perspective, the economy cannot be understood or managed outside of the social and institutional relationships that surround it, nor be extended beyond the biophysical limits imposed by ecological systems (González et al. 2008, 60). Therefore, the human system is not sustainable unless the biosphere's regeneration and load capacity are taken into account. To put it another way, the planet's biophysical limits must not be exceeded.[4]

The conceptual framework for the MA places human well-being as the central focus for assessment, while also recognizing that biodiversity and ecosystems have intrinsic value (Millennium Ecosystem Assessment 2003 and 2005b, chap. 1). Thus, the MA closely examines the concepts of ecosystem functions and services,[5] including their relationship with the components of human well-being (figure 2.1), given that changes in ecosystem services affect well-being in many ways. Figure 2.1 depicts the strength of linkages between categories of ecosystem services and components of human well-being that are commonly encountered and includes indications of the extent to which it is possible for socioeconomic factors to mediate the linkage. The strength of the linkages and the potential for mediation differs in various ecosystems and regions. In addition to the influence of ecosystem services on human well-being depicted here, other factors—including other environmental factors as well as economic, social, technological, and cultural factors—influence human well-being, and ecosystems are in turn affected by changes in human well-being (Millennium Ecosystem Assessment 2005a, 2005b).

We are often unaware of this relationship until adverse effects are produced that show how the loss of the ecosystems health is closely related to decreased human well-being. The negative impacts of water, soil and air contamination on human health are clearly among the adverse effects caused by environmental degradation and poor waste management practices. In addition, agriculture suffers due to soil degradation, which results in nutritional problems for a large part of the world's population that relies on subsistence

4. For more information on the planet's limits, see Meadows et al. (1972); Meadows, Randers, and Meadows (2004).

5. "Ecosystem functions" are defined as the capacity of the processes and structure of the ecosystems to generate services that create human well-being (de Groot 1992). "Ecosystem services" are the benefits provided by ecosystems that contribute to making human life both possible and worth living (Díaz et al. 2006, box 1), worthy of being lived. Therefore, an ecosystem function may or may not become an ecosystem service, depending on whether it contributes (directly or indirectly) to human well-being.

Figure 2.1. Linkages between ecosystem services and human well-being.

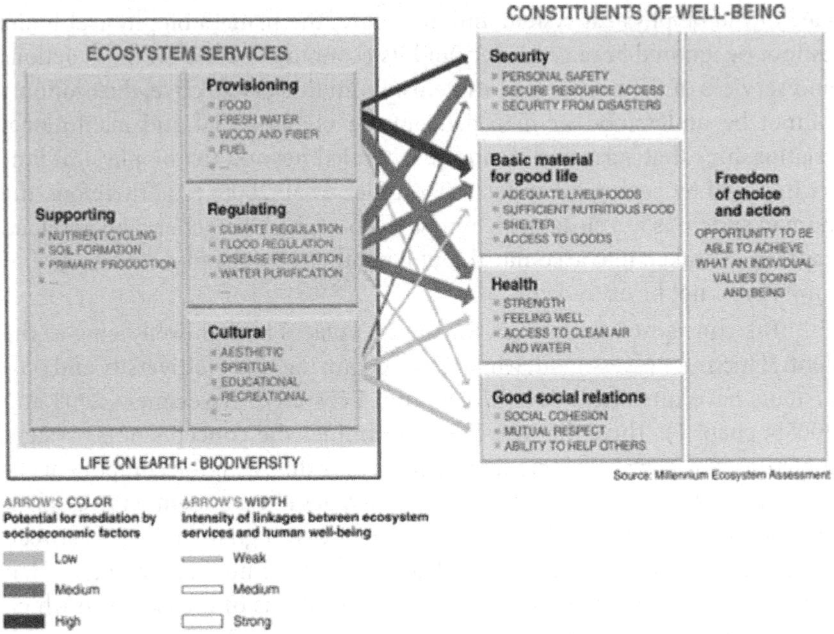

Source: Millennium Ecosystem Assessment (2005c, 24).

agriculture. Furthermore, the environment's shifts and imbalances involve major social changes and problems because they affect every type of social activity and reduce the quality of life for a large percentage of the world's population. On that point, however, the social consequences of an environmental change vary according to the specific characteristics of different societies (Duarte et al. 2006, 105), human attitudes and actions, and the capacity of societies to anticipate and make changes. This is because while well-managed ecosystems reduce risks and vulnerability, poorly managed systems can exacerbate problems by increasing risks of flood, drought, crop failure, or disease (Millennium Ecosystem Assessment 2005b, 27). For this reason, the MA underscored the need for coordinated efforts between different government sectors, researchers, companies, citizens' groups, international institutions, and other agents involved in directing the policies as well as the individual and collective actions taken to better protect natural capital.

The theoretical framework for ecosystem functions and services establish a bridge between scientific knowledge and political decision making. Through the new terminology and philosophy underlying the MA, the existing dichotomy between conservation and development has been overcome and they have ceased to be seen as two remote and conflicting issues. The importance of sustainability for well-being has begun to be seen, where conservation and the appropriate use of ecosystems have become essential elements for sustainable human development and, consequently, for human well-being.[6]

Importance of the Sub-Global Assessments in "Developed Countries"

In the case of the most industrialized, urbanized, and technologically advanced countries, it is more difficult to see the existing relationship between society and nature a priori because there are various spatial and/or temporal scalar decouplings among the place and/or moments in which the capacity to provide a service (function) is generated and in which the service of the ecosystems is taken advantage of, utilized, or enjoyed.[7] That is, residents in a large city can use services produced in distant ecosystems (e.g., food produced thousands of miles away) or in the past (e.g., in the case of oil), without the citizens being aware of the origins of those services.

However, these same societies are the ones that make greater use of the services or benefits that humans obtain from the ecosystems and from their associated functions. In Euskadi's (the Basque Country's) case, for example, the annual electricity consumption is 6,346.33 kWh./resident (Eustat, Udalmap), the annual trash production is 552kg/person, and nonhazardous waste created in the Bizkaia province is 1,728,941 tons/year (Eustat). On this point it is important to remember that environmental consumption not only marks the relationship between human beings and nature, but also the relationship among humans themselves (Sachs and Santarius 2007, 28).

For this reason, it is vitally important to use indicators that permit society and decision makers to see how industrialized and urbanized societies really use ecosystem services, near and far. To do so, a highly utilized sustainability indicator of great value in visualizing our impact on the planet and possessing a highly educative character due to its graphic overview was called for. This is the ecological "footprint," which measures "humanity's demand on the biosphere in terms of the area of biologically productive land and sea required to provide the resources we use and to absorb our waste" (World Wildlife Fund, Zoological Society of London Living Conservation, and the Global Footprint Network 2008, 14). In other words, the ecological footprint measures the required amount of useful soil that human activities consume

---

6. *Human well-being* is defined as: freedom of choice and action, in terms of meeting the needs for: (1) basic material for a good life, (2) health, (3) good social relation, and (4) security (Millennium Ecosystem Assessment 2003). The constituents of well-being, as experienced and perceived by people, are situation dependent, reflecting local geography, culture, and ecological circumstances (Millennium Ecosystem Assessment 2005b, chap. 1).

7. For more information on scalar decouplings, see Fisher, Turner, and Morling (2009).

and compares it with the surface area at our disposal. In spite of being an index that does not examine many of the environmental impacts that human beings are creating, it is sufficient to reveal that the biocapacity, or load capacity, of the planet is being exceeded. Through this and other indicators, it is easy to see how much the current societies mistakenly called "developed"[8] depend on nature and the ecosystems services. Furthermore, analyzing trends derived from these indicators over time can help track change, which is linked to certain adopted decisions, strategies, and policies that favor or harm sustainability achievements.

Thanks to these types of indicators, it is possible to recognize the impacts that the society of a particular region can create both on that region's ecosystems as well as on distant ecosystems, and thus on other geographically remote communities, are recognized. The tools to evaluate the changes created by new ways of doing things will greatly aid both local and global strategies for reducing environmental impacts, while addressing poverty, and increasing human well-being.

Millennium Ecosystem Assessment Follow-up

Given the MA's dynamic and transformative character, it is currently conducting the follow-up to the Millennium Ecosystem Assessment through a wide range of sub-global assessments (SGAs) that are coordinated by the Sub-Global Working Group to promote the implementation of the findings and recommendations of the MA .[9]

This program covers three main objectives:

1. *Building and Maintaining the Knowledge Base*—to continue to build and improve the knowledge base on the links between biodiversity, ecosystem functioning, ecosystem services, and human well-being and to develop tools for mainstreaming ecosystem services into development and economic decision making.
2. *Policy Implementation*—to strengthen country capacities to operationalize methods and tools for integrating the MA approach and its findings and recommendations in national development

---

8. Calling some societies "developed" implies that other societies are "underdeveloped." Given that there are very diverse ways of developing, and that the currently predominate development model has proven to be unsustainable, different terminology is recommended. In particular, the terminology proposed by Marcellesi and Palacios (2008, 9) is noteworthy due to its considerable potential for change. Based on this terminology, on one hand there would be industrialized countries working toward sustainable development, and on the other hand there would be countries that were not as industrialized with the potential to move directly toward sustainable human development.

9. See www.MAweb.org for further details on the MA and the various follow-up activities currently under way. See also http://devnew.gpa.nl.eu.org/ma_followup/index.php for specific information on the MA Implementation Process.

planning and policy implementation processes, including Millennium Development Goals and poverty reduction strategies at the national level.

3. *Outreach*—to disseminate the findings of the MA and its conceptual framework, tools, and methodologies to relevant stakeholders through the development of media strategies and educational tools.

The multidisciplinary character of the program is reinforced in the MA's implementation process, which, as with the global assessment, promotes the interconnection among the sub-global, local, national, and regional assessments through working groups, publications, and other work tools that help in the implementation of the assessments and favor the systemization of methodologies (see, for example, Ash et al. 2010).

The follow-up program attempts to learn from the possible mistakes and successes of MA's first phase, and, above all, strengthen the weak areas identified and enable the continuous improvement of successive assessments. To do so, common issues of SGAs concerning credibility, relevance, and legitimacy were analyzed at the annual meeting of the sub-global assessments in Bali (Indonesia) in 2010. Among others, the importance of the following aspects were highlighted: (1) stakeholder engagement and participation in the process; (2) outreach and communication; and (3) institutional strengthening by involvement in the assessment process both by "policy makers" and "policy shapers." The participation and involvement of the policy makers was important as well as that of the administration's technical experts who, on the one hand, generally remained at their job posts longer and who, on the other hand, influenced the policy makers. The need for local studies was also emphasized because case studies that could serve as examples for other regions were lacking, especially for those regions with similar characteristics.

The case study that follows is for the Bizkaia province, which is located in the northern Iberian Peninsula (43°07'N, 2°51'W) in the Basque Country. Bizkaia's assessment began in 2008 under the name *Evaluación de los Ecosistemas del Milenio en Bizkaia* (EEMBizkaia; Millennium Ecosystem Assessment in Bizkaia) and was officially accepted as an SGA for the MA implementation process on December 1, 2009, by the co-chairs of Sub-Global Working Group.

Bizkaia, which occupies a strip between the coastline and the watershed that separates the Atlantic Ocean and the Mediterranean Sea, has an expanse of 2,217 km^2 and a population of 1,151,113 people (density of 519 people/km^2) (Eustat). Its territory as a whole displays a very humanized landscape. The industrial activity, urban settlements distributed throughout all of the territory, infrastructures that link them to and provide them with other community services, and intensive productive uses established in the rural environment during the last few decades have deeply transformed the original landscape.[10]

The economic development of Bizkaia in the first stages of the twentieth century was based on the extraction of iron, which led to an iron-based industry (from mining to shipyards) that characterized the social and economic development of the province until the 1980s and 1990s. This left the territory with a high population density in the industrialized river estuaries, especially concentrated in the Greater Bilbao metropolitan area (with a density of 2,198 people/km^2) (Eustat).

Bizkaia is very heterogeneous and includes some zones of great ecological interest, such as the Gorbea and Urkiola Natural Parks. Following the economic crisis that Spain suffered at the beginning of the 1990s, the Basque Country (particularly Bizkaia) submerged itself into a deep process of transforming its secondary production sector (mainly metallurgy) into its tertiary, with Basque institutions pledging to strengthen the environment and work toward sustainability.

Bizkaia is thus a very interesting research area due to its industrial past as well as its potential for moving toward greater sustainability. Furthermore, it is noteworthy that a large amount of social, economic, and environmental information exists in both Bizkaia and the Basque Country that facilitates assessment work, which aims to collect, synthesize, and interpret the existing documentation.

The Bizkaia Case Study: Research Project Methodology and Relevant Characteristics

EEMBizkaia is sponsored by the Provincial Council of Bizkaia and scientifically coordinated by the UNESCO Chair on Sustainable Development and Environmental Education of the Universidad del País Vasco/Euskal Herriko Unibertsitatea (UPV/EHU, University of the Basque Country). The Social-Ecological Systems Laboratory from the Autonomous University of Madrid and the UNESCO Etxea Center of the Basque Country are also active participants. Data collection participation, however, is not limited to these entities. The interaction and involvement of all agents that work in the territory's macro-ecosystem is also sought.

One of EEMBizkaia's most emphasized characteristics, which distinguishes it from many other SGAs, is the strong involvement from the

10. Forest plantations currently make up 44.86 percent of the territory, with the great majority being *Pinus radiata* plantations (Gobierno Vasco 2005), which have replaced crops, meadows, and indigenous leafy hardwood forests 1995).

beginning of policy makers, managers, and technical experts from the local administration, which as the project advances manages to involve a greater number of policy makers and policy shapers. Indeed, the Provincial Council of Bizkaia is very conscious of the importance of forging ties between science and management. And so with the objective of equipping themselves with management tools based on better knowledge of the natural values of the historical Bizkaia region and the social environment, they decided to promote and get involved with this project.

This research project, which applies the MA developed methodology, works in a multidisciplinary and participatory way toward improving environmental knowledge and integration, while at the same time insisting on public environmental awareness and education. Its overall objective is to create applicable scientific knowledge for the public and private sectors regarding the possible consequences that changes in the Bizkaian ecosystems might have for society and its population's well-being, as well as to present possible response options. This is all supported in the studies and participatory processes that gather the perceptions, concerns, and contributions of the various social agents involved.

Also of note is the pursuit of stakeholder involvement in the project through participatory processes (see Ash et al. 2010, chap. 2). Thus, EEMBizkaia, through a combination of several participatory methods—surveys, in-depth interviews, and workshops—obtains information of a different type that enriches the investigation and allows the Bizkaian social-ecological systems to come together. The methods integrate the ecological, sociocultural, and economic perspectives of the human being in nature (Anderies, Janssen, and Ostrom 2004).

The workshops acquire special relevance within the participatory methodologies because they are meeting spaces for key actors and experts from very diverse areas (public administration, universities, research centers, social groups, business groups, etc.). The first of these participatory workshops was held in 2009. The central idea that emerged from the scenarios was to examine multiple possible futures and to allow their main differences to illustrate to society and to decision makers the consequences for ecosystem services and the consequences of choosing different paths on human well-being (Millennium Ecosystem Assessment 2005c).

EEMBizkaia Framework and Methodology

The study is being carried out on two distinct spatial scales: a general scale for Bizkaia and a more specific scale for different types of ecosystems and landscapes. In addition, the Bizkaia project and that of the Spanish assessment

(EME)[11] are interconnected in such a way that work is conducted jointly for the unification of methodologies and criteria. Furthermore, the Bizkaia SGA will be incorporated into the state assessment as a case study. Finally, the project forms part of the MA implementation process and as such it emphasizes the multiscale character of the project.

With regard to the scale for Bizkaia, the project studied the province's ecological footprint in 2007 based on the different ecosystems and/or types of existing surfaces in the territory. Forest masses were also studied: their condition, socio-environmental effect, and the functions and services associated with them.

On the local scale, four zones were selected within Bizkaia for the pilot study with the intention of encompassing the greatest variability possible based on the following criteria:

- Diversity of ecosystems and variety of ecosystem services
- Comparison of urban areas of high demographic pressure with rural areas
- Comparison of original, recuperated, and degraded zones
- Acknowledgment of a distribution in the entire Bizkaia territory

One of the selected areas was the greenbelt of metropolitan Bilbao, which includes Bilbao and twenty-eight other municipalities that are in the Bilbao (or Abra) Estuary (figure 2.2). These studies have been carried out through analyses of various information sources (Eustat, Udalmap, geographic information databases, earlier scientific studies, and many more) for both the Basque Country and the Bizkaia province; comprehensive background documents and scientific articles; geographic information systems; specific field studies (i.e., 450 surveys and in-depth interviews conducted in the greenbelt of metropolitan Bilbao); and contrasting participatory processes (workshops), among others.

Principal Results of the Project

The following sections discuss the principal results obtained from both the ecological footprint studies for all of Bizkaia (e.g., those gathered in the metropolitan Bilbao green belt pilot zone) and from the forest systems, as well as their interrelationship with the results from the various participatory processes carried out in the project's framework.

11. For more information on the MA in Spain, see www.ecomilenio.org.

Figure 2.2. Study area location. Position of Bilbao in relation to Europe and Euskadi, and the location of the pilot study area of metropolitan Bilbao inside of Bizkaia.

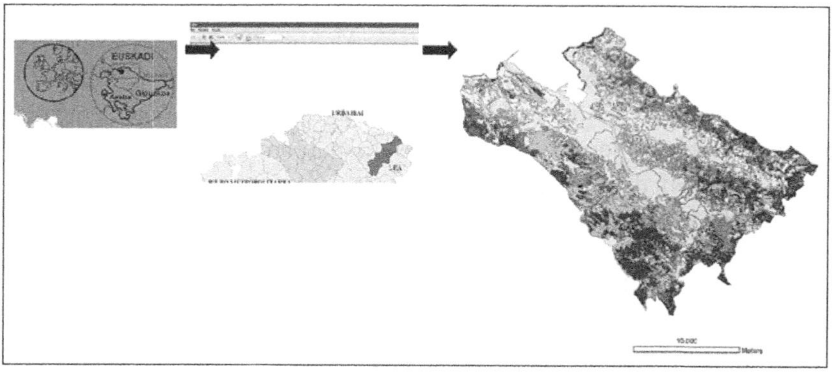

Ecological Footprint

In the principal results obtained from the ecological footprint study from Bizkaia in 2007, which is expected to have continued analysis in subsequent years, it was noted that the current way of life of Bilbao's population requires 4.84 ha/person, which is 3.83 ha/person more than what is available or 3.80 times more than what the territory can provide.

Of all the factors that determine a region's ecological footprint, the absorption area of CO_2 (determined by the production and consumption of energy, consumption of goods, and area needed for waste recovery by recycling) was the main contributor to Bizkaia's ecological footprint (4.29 ha/person). The production and consumption of energy, especially the energy incorporated into consumer goods importation, caused the most emissions and, thus, required the most area for its absorption.

Therefore, to achieve a more sustainable Bizkaia, its ecological footprint must be reduced, starting with the emissions associated with the import of consumer goods. When forming guidelines to achieve the desired future for Bizkaia, the participants in different workshop groups proposed the promotion of direct consumption, cutting out intermediaries, and minimizing transport, all of which, along with other proposed actions, would help to reduce Bizkaia's ecological footprint.

Forest Systems of Bizkaia

Bizkaia's forest systems were analyzed through a study of diverse aspects, including: environmental assessment reinforced by a characterization study; assessment of the forest landscape; socioeconomic and cultural valuation; study of the functions and services; and SWOT analysis (strengths, limitations, opportunities, threats).

In the characterization of the current forest landscape in Bizkaia it was observed that the total forest area was 57 percent of the region, of which only 12 percent corresponded to natural forests, with conifer plantations as the principal component of the landscape (representing 86,734 ha, or 69 percent forest cover). Private ownership of the total forest area was 62 percent. Oak wood ecosystems, despite making up the majority of the region's potential vegetation, actually formed an area of less than 7 percent of the total area and presented a high degree of fragmentation. In the forest landscape assessment study of the last ten years, a mixed oak forest recovery trend was observed as a result of ecological succession from scrubland. Furthermore, slight trend changes were noted in the intense forestry activity characteristic of the region, with a decline in the conifer plantations' surface area and an increase in the surface area of eucalyptus plantations, which replaced the conifer plantations and unforested zones.

In the study of the functions and services of the Bizkaia forest systems, the importance of the systems' potential functions and regulating services was observed. This point was reinforced in the workshops on forest systems, during which the participants identified many more regulating services for these systems than cultural or provisioning services. Importance was also given to the appropriate management of the forest systems so that they could maintain a high potential of offering regulating services.

For its part, the socioeconomic study clearly shows the forest sector's complicated economic situation: in the last decade the economic value of the Basque forest sector's end production suffered a 57 percent decrease, making it currently less than 0.6 percent of the Bizkaia's gross national product (Eustat).

Despite the great potential in promoting other types of activities, to date timber extraction has been the main use of the Bizkaian forests. The workshop participants recommended that the economic valuation of the forests should not only take production into account, but also other aspects, such as the tourism value that was also generating income. To achieve this, the workshop participants emphasized the need to promote and give priority to other uses of the forests besides production, through public administrations and other types of initiatives.

With the SWOT analysis, which was developed after analyzing the previous studies and reworked by the forest workshop participants, trade-off analyses acquired special relevance due to competing interests in the forest's socio-ecological systems. This fact, in turn, was confirmed in the scenarios workshops that subsequently took place, during which the forest systems were given great importance for the future of Bizkaia.

Greenbelt of Metropolitan Bilbao

The most noteworthy results and conclusions from the case study of the greenbelt of metropolitan Bilbao were as follows:

- The value attributed to ecosystem services depended on the groups of people who would benefit from them. The individual respondents' knowledge of the areas and their environmental behavior influenced their responses. The people with greater knowledge and pro-environmental behavior were those who most valued the services provided.
- The local population perceived the importance of ecosystem services for their well-being.
- The people surveyed did not identify provisioning services as important ecosystem services for the region because they did not associate those services with themselves. This was because the agricultural and livestock sector was underrepresented in the region; provisioning services came from outside. This point was reinforced as part of the participation workshop, during which it was confirmed that the order of preference of ecosystem services that the participants perceived coincided with the respondents. It was of great importance that the results obtained by two differentiated participatory methods coincided.

Furthermore, users of the metropolitan Bilbao greenbelt seemed to know that provisioning services were obtained from other ecosystems, whether near or far, demonstrating that society could sense the existing scalar decoupling. This, along with the ecological footprint results, affirmed the importance of conveying the results from these types of research projects to society to make clear that which they seemed to sense *a priori*.

Conclusions

It is important to analyze the stakeholders' perceptions of the ecosystems that they make use of, as well as to create ties between research, management, and the citizens through participatory processes that aid in broadening knowledge and improving the management of the social–ecological systems. In cases where conflicts of interest could occur, a deeper examination of the studies and a search for common workspaces that encourage consensus and creativity is recommended, in the pursuit of new paths toward sustainability.

References

Anderies, John M., Marco A. Janssen, and Elinor Ostrom. 2004. "A Framework to Analyze the Robustness of Social-Ecological Systems from an Institutional Perspective." *Ecology and Society* 9, no. 1: 18.

Ash, Neville, Hernán Blanco, Claire Brown, Keisha Garcia, Thomas Henrichs, Nicolas Lucas, Ciara Raudsepp-Hearne, R. David Simpson, Robert Scholes, Thomas Tomich, Bhaskar Vira, and Monika Zurek. 2010. *Ecosystems and Human Well-being: A Manual for Assessment Practitioners*. Washington, D.C.: Island Press.

Berkes, Fikret, Johan Colding, and Carl Folke. 2003. "Introduction." In *Navigating Social-Ecological Systems: Building Resilience for Complexity and Change*, edited by Fikret Berkes, Johan Colding, and Carl Folke. Cambridge: Cambridge University Press.

Daily, Gretchen C. 1997. *Nature's Services: Societal Dependence on Natural Ecosystems*. Washington, D.C.: Island Press.

de Groot, Rudolph S. 1992. *Functions of Nature: Evaluation of Nature in Environmental Planning, Management and Decision Making*. Groningen, Netherlands: Wolters-Noordhoff.

Díaz, Sandra, Joseph Fargione, F. Stuart Chapin, and David Tilman. 2006. "Biodiversity Loss Threatens Human Well-being." *PLoS Biology* 4, no. 8: 1300–1305.

Duarte, Carlos M., Sergio Alonso, Gerardo Benito, Jordi Dachs, Carlos Montes, Mercedes Pard, Aida de la Aurora Fernández Ríos, Rafael Simó, and Fernando Valladares. 2006. *Cambio global: Impacto de la actividad humana sobre el sistema Tierra*. Madrid: Consejo Superior de Investigaciones Científicas (CSIC).

Eustat: Basque Statistics Institute. At http://en.eustat.es/idioma_i/indice.html#axzz1vWYTKqRH.

———. Udalmap: Municipal-Level Indicators of Sustainability. At http://en.eustat.es/ci_ci/about/udalmap_i.html#axzz1vWYTKqRH.

Fisher, Brendan, Kerry R. Turner, and Paul Morling. 2009. "Defining and Classifying Ecosystem Services for Decision Making." *Ecological Economics* 68, no. 3: 643–53.

Gobierno Vasco. 2005. "Inventario Forestal CAE." Departamento de Medio Ambiente, Planificación Territorial, Agricultura y Pesca del Gobierno Vasco. At www.nasdap.ejgv.euskadi.net/r50-15135/es/contenidos/informacion/inventario_forestal_index/es_dapa/inventario_forestal_index.html.

González, José Antonio, Carlos Montes, Ignacio Santos, and C. Monedero. 2008. "Invirtiendo en capital natural: Un marco para integrar la sostenibilidad ambiental en las políticas de cooperación." *Ecosistemas* 17, no. 2: 52–69.

Marcellesi, Florent, and Igone Palacios. 2008. *Integración de consideraciones de sostenibilidad en la cooperación para el desarrollo*. Cuadernos Bakeaz 88. Bilbao: Bakeaz.

Martín-López, Berta, Erik Gómez-Baggethun, José A. González, Pedro L. Lomas, and Carlos Montes. 2009. "The Assessment of Ecosystem Services Provided by Biodiversity: Rethinking Concepts and Research Needs." In *Handbook of Nature Conservation: Global, Environmental and Economic Issues*, edited by Jason B. Aronoff. New York: Nova Science Publishers.

Meadows, Donella H., Dennis L. Meadows, and Jørgen Randers. 1992. *Beyond the Limits: Confronting Global Collapse Envisioning a Sustainable Future*. Post Mills, VT: Chelsea Green Publishing Company.

Meadows, Donella H., Dennis L. Meadows, Jørgen Randers, and William Behe-rens. 1972. *The Limits to Growth: A Report for the Club of Rome's Project on the Predicament of Mankind*. New York: Universe Books.

Meadows, Donella H., Jørgen Randers, and Dennis L. Meadows. 2004. *Limits to Growth: The 30-Year Update*. White River Junction, VT: Chelsea Green Publishing Company.

Millennium Ecosystem Assessment. 2003. *Ecosystems and Human Well-being: A Framework for Assessment*. Washington, D.C.: Island Press. At http://www.millenniumassessment.org/en/Framework.aspx.

——. 2005a. *Ecosystems and Human Well-being: Synthesis*. Washington, D.C.: Island Press. At www.millenniumassessment.org/documents/document.356.aspx.pdf.

——. 2005b. *Ecosystems and Human Well-being: Current State and Trends*. Washington, D.C.: Island Press. At www.millenniumassessment.org/en/Condition.aspx.

——. 2005c. *Ecosystems and Human Well-being: Scenarios*. Washington, D.C.: Island Press. At www.millenniumassessment.org/en/Scenarios.aspx.

Onaindia Olalde, Miren. 2007. "Sostenibilidad Ecológica." *Forum de Sostenibilidad* 1: 39–49.

Rescia, Alejandro J., Ma. Fe Schmitz, Ma. Pilar Martín de Agar, Carlos L. de Pablo, and Francisco D. Pineda. 1995. "Ascribing Plant Diversity Values to Historical Changes in Landscape: A Methodological Approach." *Landscape and Urban Planning* 31, nos. 1–3: 181–94.

Sachs, Wolfgang, and Tilman Santarius, eds. 2007. *Fair Future: Resource Conflicts, Security and Global Justice: A Report of the Wuppertal Institute for Climate, Environment and Energy*. Translated by Patrick Camiller. London and New York: Zed Books; Halifax, Nova Scotia: Fernwood; New York: Distributed in the USA exclusively by Palgrave Macmillian.

Wackernagel, Mathis, and William E. Rees. 1996. *Our Ecological Footprint: Reducing Human Impact on Earth*. Gabriola Island, BC, Canada: New Society Publishers.

World Wildlife Fund (WWF), Zoological Society of London Living Conservation, and Global Footprint Network. 2008. *Living Planet Report 2008*. At http://assets.panda.org/downloads/lpr_2008_1.pdf.

3

Climate Change: Activities of the EOLO Group at the University of the Basque Country

Agustín Ezcurra, Jon Sáenz, and Gabriel Ibarra-Berastegi
The University of the Basque Country (UPV/EHU)

World population growth has continued to explode since the start of the industrial revolution. As such, massive scale emissions of atmospheric gases and particles from combined human activities have been unceasing and humankind has now left a noticeable mark on the chemical composition of the atmosphere. The current concentration of carbon dioxide on the planet is the highest it has been in at least 650,000 years and primarily comes from the burning of fossil fuels carried out by man.

This imbalance is the result of a systemic increase in per person lifetime average energy consumption as well as the continual increase of the world's population and average life expectancy. Given the inefficiency of modern technologies, the demand for increasing amounts of energy implies that the natural environment that surrounds us will be contaminated. As the number of individuals benefiting from the improvements inherent to a technological civilization continually rises, the contamination will increase daily. It should be obvious to us that the particular contamination we produce comes from the specific processes we employ to create the goods we use and the energy we consume.

There is no scientific debate whatsoever on this issue. In order to fuel the ongoing industrialization process that began at the end of the eighteenth century, humans have injected enormous amounts of carbon dioxide into the atmosphere, thus overexploiting fossil fuels that are now used daily and on a massive scale. The composition of the atmosphere is currently changing at an extremely fast rate due to the consumption levels of developed countries and the arrival of emerging industrial powers such as India, Brazil, and China, where over half of the human population exists. Within a few years, we will find ourselves on a planet with more than 9 billion people, nearly half of whom will be living in industrial areas and large cities and consuming enormous amounts of energy. Without the development of currently unknown technologies, energy consumption will continue to come specifically from the burning of fossil fuels.

Many have foreseen and given warning to humankind's movement toward an industrialization of the entire planet. This scenario has been associated with damaging effects on the environment and mankind itself. At the same time, this problem has become a political issue whose primary emissaries in recent history have been environmental movements. In this regard, in any report on the history of modern environmental thinking, it is well documented that the 1972 United Nations (UN) Conference held in Stockholm, Sweden, marked the first expression of global concern for the environment. Since then, environmentalism has gradually come to form part of a citizen consciousness and is now a reality that cannot be left out of political discourse or discourse within the business world.

Climate Change: Science and Politics

Within the general debate, in the last few years a discussion has begun regarding the possibility of climate change associated with the industrialization process. Some of the substances generated by human activity that are emitted into the atmosphere have the special characteristic of being able to produce what is known as the *greenhouse effect*. This effect, which occurs naturally in the environment, is when certain gases absorb considerable amounts of thermal radiation from the planetary surface and then return it back toward the surface. Therefore, humankind's footprint on the planet becomes noticeable as the population and its prosperity increase. This not only produces an imbalance in the natural geochemistry of the atmosphere, but also a change in its radiant energy flux.

Because meteorological phenomena and climate are nothing more than expressions of the thermal processes they produce in the atmosphere, it can be deduced that local and global meteorology and planetary climatology are affected by the presence of humankind. This assertion is not a catastrophic speculation, but rather a fact derived from observation of the chemical composition of the atmosphere and the fact that humans are modifying that composition in a very real and specific manner.

This is not a new or particularly surprising conclusion. For example, all scientists accept as fact that the proliferation of plants and other beings capable of photosynthesis on the planet has modified the chemical composition of the atmosphere and, as a result, the global climate. It is also a fact that in today's cities there is an obvious local climate change occurring that is found to be more acute when a city has a higher population. Based on the existing fossil record, we know that the climate has varied since the beginning of earth's existence. All of these facts have been extensively analyzed in current climatology studies.

Given the importance of this issue, for several decades the global scientific community has been committed to researching the potential direction that human-produced climate change may take. Throughout the world, groups of specialists in meteorological observation and measurement as well as those who model atmospheric, oceanic, and biochemical phenomena have been working to attempt to answer the question of the potential direction of human-produced climate change. Given the importance of this issue, these studies have been sponsored and coordinated by the UN under the wing of the Intergovernmental Panel on Climate Change (IPCC).

The results of these studies show that climate change caused by humans involves an increase in the average temperature of Earth's atmosphere and possibly an intensification of some meteorological phenomena in various regions, including, for example, precipitation and heat waves. These changes must be compared to the natural probability of their occurrence without the presence of —and we have already seen— an increasingly polluting human population. Obviously, it cannot be proven that every heat wave is the result of climate change. But it can be stated that there is a higher probability that this type of phenomena will occur.

In such a context, there are no scientists who doubt that the presence of varying amounts of carbon dioxide gas in the atmosphere has exerted a deep influence on climatic variation, along with other factors, including continental drift and orbital variation, among others. The gas affects the climate because it absorbs thermal infrared radiation emitted by Earth and therefore warms the atmosphere where it is found. The atmosphere then re-emits infrared radiation, thereby increasing the planet's surface temperature.

It has taken scientists nearly a century to determine with a margin of confidence that an increase in the concentration of carbon dioxide can produce a warming of the planetary surface. Beginning with the first studies completed by Joseph Louis Fourier and John Tyndall in the middle of the nineteenth century until the Fourth Assessment Report by the IPCC (2007), experimental techniques have been refined and have even led to the development of remote measurement techniques using artificial satellites.

Experiments in measurement and observation in conjunction with emerging technologies have allowed more complex computational systems to be constructed, thus allowing: simulation of the physical mechanisms implied in the climate's response to external anthropogenic forces; weather prediction; and determination of the potential condition of an ocean in any part of the world.

Policy makers have supported studies of the earth that require mutlidisciplinary and multinational collaboration in order to measure the effects of advancing globalization. Beginning with the first International Geophysical Year (1957) and continuing through the more recent International Polar Year (2007–2009), various experiments have been designed and coordinated on a global level.

In some ways, the work of the IPCC has been nothing more than the natural consequence of fundamental developments in scientific knowledge produced in physics, mathematics, and technology in the past 150 years. Therefore, a majority of scientists and politicians accept that the work of the IPCC constitutes the best estimates available concerning the problem of global warming.

There is, obviously, a level of uncertainty in the results obtained. However, the most significant uncertainties are not related to the physical or chemical processes involved in the problem. The specific results for desired future modifications will change according to the hypotheses implemented in their procurement, hypotheses that affect the IPCC's *scenarios*. These scenarios are a combination of demographic assessments, economic development, and potential and likely future human energy sources. The emissions scenarios simply attempt to provide a margin of error between more or less consumerist hypotheses (cleaner or more obsolete technologies) or more or less controlled demographics. Therefore, these projections for the future are simply *ranges* that attempt to define future situations regarding greenhouse gas emissions. Apart from these issues, a majority of the scientific community agrees on the most common elements presented in the results offered by the different working groups.

There are, however, individuals who have voiced disagreements with some of the specific results that the IPCC researchers have obtained. This does not imply that the studies performed by the numerous scientists who have been working on this problem for many years have been manipulated in any way. In order to advance, science needs skeptical critics because they are sometimes able to point out inconsistencies in certain reasoning and certain observation and calculation procedures. The group of critics who deny the existence of anthropogenic climate change is growing smaller. Members of this remaining group have not been able to provide convincing evidence that human actions are not altering the climate.

Nevertheless, in recent years this problem has been approached in a markedly partisan manner, most likely due to the financial interests at stake. If a scientist maintains that the IPCC's conclusions (including uncertainties) are reasonable, that balance of current knowledge, that scientist runs the serious risk of being accused in certain media circles of being an irresponsible environmentalist. Thus, a false public debate issue has been created about something that is simply the result of the efforts of hundreds of scientists around the world working for almost a century in ongoing research investigations.

We find ourselves faced with a situation similar to the one that occurred with tobacco consumption. Now, as at that time, the media and journalists are very quick to question scientific results. These critiques come from obvious stakeholders or simply from a position of complete lack of knowledge on the issue. However, these opinions are common, given the urgent obligation to say something about the topic in the media. What is certain is that some isolated errors detected in the reports issued by IPCC Working Groups 2 and 3 have not helped to create a fair perception of the quality of scientific study performed overall. Neither has the idea that "the Earth is warming," an expression that in no way describes the average statistical trend predicted for the specific and determined future through particular socioeconomic scenarios.

In any case, very important decisions are being made based upon the current widespread ecological thinking among young people and adults. For example, Spain's most recent governments have upheld a popular nuclear moratorium and support alternative energy production with tax subsidies for wind and solar power.

In other words, even if there were no anthropogenic climate change, the idea of full-out development based on continued consumption of fossil fuels would not be well received by the general public who, by means of their votes, sustain the politicians who govern them. For a majority of people, environmentalism now forms part of their thinking. As such, in developed countries no one is prepared to imagine that we are living on a planet where nature is shrinking around us and is being devoured by uncontrolled industrial advance. Within this scenario of global environmentalism, climate change is another truth added to the rest of the problems affecting citizens. This truth is inconvenient but, unfortunately, it is also real. In facing this fact, some professional critics from the media have reacted thoughtlessly, *trying to shoot the messenger* who bears the message. In other words, they reject reports from scientists who, based on observed facts and the outcomes of the theories that explain them, have informed us that Earth's climate is being altered by human activities.

Climate Change: Situation in the Basque Country

Because the topic of discussion here is climate and climate change, the general characteristics that determine the Basque Country's climatology may be interesting to consider. Latitude is probably the geographical factor that most determines the climate of the Basque Country, which is also called Euskadi (in the Basque language, Euskara). The inclination of the sun's rays during different seasons of the year depends on latitude. The Basque Country's latitude, between 42° and 43.5° north of the equator, places the country within what has been called the *temperate zone.*

Another important element that affects the Basque Country's climatology is the Gulf Stream. Because of this oceanic current, the coasts of Europe enjoy a much more mild winter than would correspond to their latitude. If not for this system of ocean currents from the Atlantic, Euskadi's winter climate would be quite different: much colder and, most likely, with much less precipitation. Other oceanic influences that hold certain importance in the Basque climate are those coming from the Mediterranean Sea and also the Cantabrian Sea, which has a direct and significant influence on the climate in the northern slope of the Basque Country.

The landforms of the Basque Country create notable differences in the climatic parameters of its various orographic zones. The Iberian Peninsula's land relief also influences the general characteristics of the Basque climate. The hilly nature of the peninsula, which has a significant average altitude (660 m) and multiple mountainous obstacles, causes the Atlantic southeast flows to arrive at Euskadi with characteristics that have been significantly altered by the Föhn effect (a type of dry down-slope wind that occurs in the lee or downwind side of a mountain range).

The Basque Country's climate is divided into different regional micro climates due to the effects of mountain ranges. Valleys in the north—in Bizkaia, Gipuzkoa (Guipúzcoa), and the Ayala Valley in Araba (Álava)—are characterized by a predominating oceanic climate with constant humidity and moderate temperatures. The average annual precipitation is close to 1200 mm, with the heaviest rains occurring from October to December. Temperatures vary between 25°C and 14°C in summer and do not drop below 5°C in winter.

Central and southern regions have a Continental Mediterranean climate but are also influenced by the Northern Oceanic. This creates warm, dry summers and cold winters with snowfall. Rainfall is scarce and irregular. Precipitation tends to reach its peak in spring and fall, while the annual average can be as low as 300 mm in some regions. In winter the temperature hovers at around 4 C amidst days when it may fall as low as -9.5 C. Summers have an average temperature of 20°C with highs up to 35 C.

Finally, possible fluctuations in the Basque Country's climate due to the greenhouse effect are linked to those observed in the climate of the Iberian Peninsula. Their primary causes are basically determined based on its geographical position in the extreme southeast of the Euro-Asian continent.

This unavoidably links these fluctuations to the changes that may occur in the positioning of the Polar Jet Stream (Ulbrich et al. 1999; Sáenz, Zubillaga, and Rodríguez-Puebla 2001); the permanent Azores Anticyclone (Sáenz et al. 2001); and the Mediterranean Front.

Studies on the climate change trends occurring in the Basque Country indicate that this phenomenon is just beginning to be felt. An increase in temperatures higher than the average global mean has become clear, although part of this trend is attributable to climate variability in the Atlantic region (Rodríguez-Puebla et al. 2010). In fact, the increase of the Basque Country's average temperature could reach 0.8°C in 2020 and 1.3°C in 2050.

Climate Change: How Do the Models Operate on a Global Scale?

Our research group, named EOLO (www.ehu.es/eolo/index_en.html), is comprised of scientists and professors from the UPV/EHU. Our research is focused on climate, meteorology, and the environment. EOLO has developed applied research on problems of climate change, and one of the most recent projects (Errasti et al. 2010) is a study of the validity of the different climate models used by the IPCC for the Iberian Peninsula region. These types of studies are fundamental when deciding which models should be used to predict future climate behaviors for a specific region. Additionally, in collaboration with other Spanish research groups, EOLO has also carried out studies on the possible strategies for regional analysis and the influence of natural variations in climate change processes.

Global climate models (GCMs) used by the IPCC are sophisticated numeric tools designed to simulate the climate and require powerful computers to be executed. Only a small number of research organizations in the world have developed these models and have integrated them in support of the IPCC's activities. However, the numeric models are not perfect and produce results with certain indeterminacies. As a result, and prior to utilization of the IPCC model results to establish future climate scenarios, the accuracy of these numeric models must first be evaluated. Specifically, when the climate models will be used as tools for managing future impacts, performance of an evaluation is a necessary first step.

Several scholarly articles address this issue (Ahlfeld 2006; Lucarini et al. 2007; Maxino et al. 2007; Nieto and Rodríguez-Puebla 2006; Reichler and Kim 2008; and Tebaldi et al. 2006). All of these authors study the operation of the different climate models, measuring their ability to describe today's climate on various levels, from global to regional. The results of these studies indicate that not all of the IPCC models are able to describe the present climate with the same accuracy and that the best models for a region may not be as good when applied to other regions. Furthermore, their accuracy varies when different meteorological variables are examined.

Climate Change: Which Model Can Best Predict Future Climate?

A model's ability to adequately reproduce current climate capacity does not necessarily guarantee the quality of its predictions for the future. But despite this consideration, it is generally accepted that the quality of the models' ability to reproduce current observations is the only way for them to be proven credible, under the assumption that a model that describes current climate with accuracy will perform a better future projection than a model that does not correctly reproduce today's climate for a specific region. The primary objective of the research that the EOLO team carried out (summarized below), was to try and select the AR4 models that would most accurately reproduce the climate in the Iberian Peninsula during the period of 1979–1998.

Nieto and Rodríguez Puebla (2006) previously performed a study that focused on validating the accuracy of nine climate models for the Iberian Peninsula. These authors tested the ability of nine AR4 models to describe the natural variability of winter and summer precipitation for the geographical area. Their study was based on an evaluation of the seasonal cycles through use of Empirical Orthogonal Function (EOF) along with Fourier spectral analysis methods.

Our work attempted to move this type of study for the Iberian Peninsula one step further by evaluating the ability of the majority of the IPCC models to describe the Iberian climate today. Thus, the study that we carried out analyzed the meteorological variables of twenty-four models developed by sixteen institutions worldwide. The data used was extracted from the data archives of the Program for Climate Model Diagnostics and Intercomparison (PCMDI). Some of these models have produced only one execution (run), while other models have up to nine different runs.

The final determined ranking would be beneficial to the community of scientists who study climate and its evolution in Spain and would allow for a more precise description of future climate scenarios for our area of interest. In our analysis, our focus was on the variables that are most typically used in downscaling exercises: sea-level air pressure, precipitation, and temperature.

Our assessment work followed a strategy different than that used by other researchers. The most deterministic part of any climatic element for a mid-latitude area of Earth is its seasonal cycle. Therefore, the models' ability to measure this portion of the climate indicators was considered. Its capacity to describe the statistical climate characteristics through a comparison of probability density functions (PDFs), a common instrument for evaluating the results from climate models, was also considered.

To assess the simulations obtained with the IPCC AR4 models for the period of comparison, their data was compared with the following real observed baseline data for the current climate:

a. Sea-level pressure (SLP) data, surface air temperature (TAS) at 10 meters, and precipitation (PR) from the ERA40 reanalysis project by the European Centre for Medium-Range Weather Forecasts, or ECMWF (Simmons and Gibson 2000; Uppala et al. 2005).
b. The combined precipitation (PR) from the Global Precipitation Climate Project (GPCP), developed by George J. Huffman et al. (1995).

To acquire a coherent comparison between the models and observed data, a common and uniform 2.5° x 2.5° latitude–longitude grid was used. Therefore, the data from the models and from the observations were preprocessed using a bilinear interpolation and were projected onto the same spatial grid. The spatial window used in performing our study was [10° E - 2.5° W, 35° - 45° N]. Within this window, when the observations and data from the climate models were ultimately compared, the Iberian Peninsula was covered by a total of 30 grid points. The seasonal rate chosen for the comparison was the rate of the monthly averages.

To compare the seasonal cycles, the correlation coefficient (r) and the root-mean-square error (rms) obtained were used as a metric to compare the modeled seasonal cycles with the observed seasonal cycles. To compare the PDFs (PDFs for modeled data against PDFs for observations) at each point in the Iberian Peninsula, the "s" parameter (called a *skill score*), previously used by Claire C. Maxino et al. (2007), was used. This parameter provides a simple but useful measurement of similarity between two PDFs.

The results of the comparison based on the seasonal cycle showed that both the analysis of the correlation coefficient "r" and the "rms" generate very similar classifications between the models. However, it could be observed that the "rms" test was more selective than the "r" test.

The study of the comparison of the PDFs ranked the IPCC models in a manner similar to the first statistical results previously obtained, with the exception that in some cases the skill score provided a better classification than the comparison of the seasonal cycles. In other words, although the statistical distribution of some models reproduces current observations, that fact does not guarantee that the seasonal cycle will be exactly reproduced. This result indicates the importance of analyzing the seasonal cycles when validating climate models. In addition, it was also observed that the models with good results in describing a specific meteorological variable at times did not reach the same level of operation for other variables.

Through a combination of the different classifications performed, as a definitive result, the following were shown to be the best climate models to be used in the Iberian Peninsula: MIROC3.2-HIRES, GFDL-CM2.1, UKMO-HADGEM1, MIROC3.2-MEDRES, and MPI-ECHAM5.

Finally, the rankings found in this study are significant from the perspective of the area studied and the variables analyzed (sea-level pressure, temperature, and precipitation). An extension of the conclusions from this study to a general classification of the models in other areas or diagnostics could be misleading and should absolutely be avoided.

Climate Change: The Role of Natural Variations

Notwithstanding what has been mentioned regarding the fact that some trends are already evident in the seasonal cycles of elements of the Basque Country's climate, it is true that part of the variability of these cycles is due to low-frequency climatic oscillations, also known as *teleconnections*. Thus, it is known that the primary factor underlying the cause of the fluctuations in the winter temperature cycles above the zone is associated with variations in the patterns of the eastern Atlantic, something that has been identified (at least in the winter cycle) in various studies on the evolution of said winter temperature.

Thus, the fraction of total variance of the winter temperature above the area that can be explained based on changes to the enthalpy density flows due to seasonal circulations is greater than that corresponding to the north Atlantic oscillation (NAO), which is due to the position of the maximum gradient of anomalies associated to this pattern over the zone of study (Sáenz et al. 2001). This is not an isolated fact, given that the eastern Atlantic pattern also appears as an important influence on humidity flows in the Mediterranean basin (Fernández, Sáenz, and Zorita 2003); seasonal probability of the temperature above the Iberian Peninsula (Frías et al. 2005); and anchovy recruitment in the Bay of Biscay (Borja et al. 2008). Studies performed to date indicate that there are various physical mechanisms involved in changing different variables. We hope that similar future studies can further develop this progress.

Regionalization Strategies

In recent years, the development of future climate scenarios with a spatial resolution high enough so that these studies can be useful for the existing working communities in the area of impacts has been a priority of climatology research in advanced countries. In general, these increased scale studies can be classified in two approaches. The first involves integrating a regional climate model nested into the grid of a general circulation model, a technique called *downscaling*. EOLO has participated in these types of studies in collaboration with other groups (Fernández et al. 2007) using the dynamic Weather Research and Forecast model (WRF). These studies are currently being extended to the creation of a regional reanalysis using 3DVAR assimilation of observations, work that has been developed with computational support through a special ECMWF project.

The second approach uses *statistical downscaling models* to obtain regional climate scenarios. EOLA's experience with this approach is quite varied. This is due to the development of methods inherent to regionalization that allow not only the predictand fields to be included in the process, and also the predictor fields (Fernández and Sáenz 2003); these studies have recently expanded to the development of systems for predicting precipitation in the very short term (nowcasting). Both a deterministic model (Fernández-Ferrero et al. 2009) and a probabilistic model have been developed for this (Fernández-Ferrero, Sáenz, and Ibarra-Berastegi 2010) based on different Bayesian methods.

Conclusion

The climate change problem, or, more generally, the problem of adequately managing our future in environmental terms, is currently a challenge on both a political and an investigative level because most recent observations indicate the presence of an intense anthropogenic effect that, on a global and local scale, is changing the natural balances of meteorology and the climate as well as the balance of many land and marine ecosystems. Uncertainties on these issues, while they do exist, do not affect the problem's fundamental core, which is the simple fact that human activity has now reached a level where its effects are felt around the planet and not just in certain regions.

Faced with this situation, society as a whole must know how to react before we find ourselves past the point of no return. To do so, however, a close collaboration must be developed between political groups, economic groups, the media, civil society, and scientists. This collaboration cannot be initiated without an honest and ongoing dialogue between all parties who can and must participate in the solution of this problem. But above all else, it seems clear that the work of scientists who are dedicated to this purpose must be kept in mind; the rate of their work must be respected, and their results must be taken into consideration. There should be no doubt that all scientists working on this issue are doing so with honesty and that scientific investigation is the only possible route to follow in finding solutions that will mitigate the effects observed and those predicted as possible. This does not imply acceptance without replication of those scientific results obtained up to now. Science is also constructed through debate, but it must be a calm and rational debate with the goal of always eliminating the scientific uncertainties or contradictions that may be detected in the conclusions obtained.

References

Ahlfeld, David P. 2006. "Comparison of Climate Model Precipitation Forecasts with North American Observations." In *Proceedings of the XVI International Conference on Computational Methods in Water Resources*, edited by Philip J. Binning, Peter Engesgaard, Helge Dahle, George F. Pinder, and William G. Gray. Copenhagen, Denmark, June.

Borja, Ángel, Almudena Fontán, Jon Sáenz, and Victoriano Valencia. 2008. "Climate, Oceanography, and Recruitment: The Case of the Bay of Biscay Anchovy (*Engraulis encrasicolus*)." *Fisheries Oceanography* 17, no. 6: 477–93.

Errasti, Iñigo, Agustín Ezcurra, Jon Sáenz, and Gabriel Ibarra-Berastegi. 2010. "Validation of IPCC AR4 Models over the Iberian Peninsula." *Theoretical and Applied Climatology* 103, nos. 1–2: 61–79.

Fernández, Jesús, and Jon Sáenz. 2003. "Improved Field Reconstruction with the Analog Method: Searching the CCA Space." *Climate Research* 24, no. 3: 199–213.

Fernández, Jesús, Jon Sáenz, and Eduardo Zorita. 2003. "Analysis of Wintertime Atmospheric Moisture Transport and Its Variability over Southern Europe in the NCEP-Reanalysis." *Climate Research* 23, no. 3: 195–215.

Fernández, Jesús, Juan P. Montávez, Jon Sáenz, J. Fidel González-Rouco, and Eduardo Zorita. 2007. "Sensitivity of the MM5 Mesoscale Model to Physical Parameterizations for Regional Climate Studies: Monthly Seasonal Cycle." *Journal of Geophysical Research—Atmospheres* 112: D04101.

Fernández-Ferrero, Alejandro, Jon Sáenz, Gabriel Ibarra-Berastegi, and Jesús Fernández. 2009. "Evaluation of Statistical Downscaling in Short Range Precipitation Forecasting." *Atmospheric Research* 94, no. 3: 448–61.

Fernández-Ferrero, Alejandro, Jon Sáenz, and Gabriel Ibarra-Berastegi. 2010. "Comparison of the Performance of Different Analog-based Bayesian Probabilistic Precipitation Forecasts over Bilbao, Spain." *Monthly Weather Review* 138, no. 8: 3107–19.

Frías, M. Dolores, Jesús Fernández, Jon Sáenz, and Concepción Rodríguez-Puebla. 2005. "Operational Predictability of Monthly Average Maximum Temperature over the Iberian Peninsula using DEMETER Simulations and Downscaling." *Tellus A* 57, no. 3: 448–63.

Huffman, George J., Robert F. Adler, Mark M. Morrissey, David T. Bolvin, Scott Curtis, Robert Joyce, Brad McGavock, and Joel Susskind. 2001.

"Global Precipitation at One-Degree Daily Resolution from Multi-Satellite Observations." *Journal of Hydrometeorology* 2, no. 1: 36–50.

Lucarini, Valerio, Sandro Calmanti, Alessandro Dell'Aquila, Paolo M. Ruti, and Antonio Speranza. 2007. "Intercomparison of the Northern Hemisphere Winter Mid-latitude Atmospheric Variability of the IPCC Models." *Climate Dynamics* 28, nos. 7–8: 829–48.

Maxino, Claire C., Brian J. McAvaney, Andrew J. Pitman, and Sarah E. Perkins. 2008. "Ranking the AR4 Climate Models over the Murray Darling Basin using Simulated Maximum Temperature, Minimum Temperature and Precipitation." *International Journal of Climatology* 28, no. 8: 1097–1112.

Nieto, Susana, and Concepción Rodríguez-Puebla. 2006. "Comparison of Precipitation from Observed Data and General Circulation Models over the Iberian Peninsula." *Journal of Climate* 19, no. 17: 4254–75.

Reichler, Thomas, and Junsu Kim. 2008. "How Well do Coupled Models Simulate Today's Climate?" *Bulletin of the American Meteorological Society* 89, no. 3: 303–11.

Rodríguez-Puebla, Concepción, Ascensión H. Encinas, Luis Alberto García-Casado, and Susana Nieto. 2010. "Trends in Warm Days and Cold Nights over the Iberian Peninsula: Relationships to Large-scale Variables." *Climate Change* 100, nos. 3–4: 667–84.

Sáenz, Jon, Juan Zubillaga, and Concepción Rodríguez-Puebla. 2001. "Interannual Variability of Winter Precipitation in Northern Iberian Peninsula." *International Journal of Climatology* 21, no. 12: 1503–13.

Sáenz, Jon, Concepción Rodríguez-Puebla, Jesús Fernández, and Juan Zubillaga. 2001. "Interpretation of Interannual Winter Temperature Variations over Southwestern Europe." *Journal of Geophysical Research—Atmospheres* 106, no. D18: 20641–52.

Simmons, Adrian J., and J.K. Gibson. 2000. *The ERA-40 project Plan*. ERA-40 Project Report Series No. 1. Reading, UK: European Centre for Medium-Range Weather Forecasts.

Tebaldi, Claudia, Katharine Hayhoe, Julie M. Arblaster, and Gerald A. Meehl. 2006. "Going to the Extremes: An Intercomparison of Model-Simulated Historical and Future Changes in Extreme Events." *Climatic Change* 79, nos. 3–4: 185–211.

Ulbrich, Uwe, Michael Christoph, Joaquim G. Pinto, and João Corte-Real. 1999. "Dependence of Winter Precipitation over Portugal on NAO and Baroclinic Wave Activity." *International Journal of Climatology* 19, no. 4: 379–90.

Uppala, Sakari M., Per W. Kållberg, Adrian J. Simmons, Ulf Andrae, Vanda Da Costa Bechtold, Michael Fiorino, J.K. Gibson, Jan Haseler, Angeles Hernández, Graeme A. Kelly, Xiaofan Li, Kazutoshi Onogi, Sami Saarinen, Niko Sokka, Richard P. Allan, Erik Andersson, Klaus Arpe, Magdalena A. Balmaseda, Anton C.M. Beljaars, Leo van der Berg, Jean-Raymond Bidlot, Niels Bormann, Sofia Caires, Frédéric Chevallier, Anje Dethof, Milan

Dragosavac, Mike Fisher, Manuel Fuentes, Stefan Hagemann, Elias Hólm, Brian J. Hoskins, Lars Isaksen, Peter A.E.M. Janssen, Roy Jenne, Anthony P. McNally, Jean-François Mahfouf, Jean-Jacques Morcrette, Nick A. Rayner, Roger W. Saunders, P. Simon, Andreas Sterl, Kevin E. Trenberth, Agathe Untch, Drasko Vasiljevic, Pedro Viterbo, and John Woollen. 2005. "The ERA-40 Re-analysis." *Quarterly Journal of the Royal Meteorological Society* 131, no. 612: 2961–3012.

4

The Environmental Value of the Karstic Landscape of the Urdaibai Biosphere Reserve: The Asnarre Promontory (Bizkaia)

Arantza Aranburu, Laura Damas-Mollá, Patxi García-Garmilla, Iñaki Yusta, Miren Josu Arriolabengoa, Peru Iridoy
University of the Basque Country (UPV/EHU)
Eneko Iriarte
University of Cantabria

Eyes only see what they look at, and they only look at what they already know. We might add that if they do not find what they are seeking, they say there is nothing.

—Telesforo de Aranzadi

Environmental values can be understood as a combination of qualities that define a specific natural environment, including the characteristics of the live, inert, and cultural components. One of these values is landscape, a product of the interaction between the various factors (abiotic, biotic, and anthropogenic) that are present and that are reflected visually in the space. Although these three pillars comprise the study of landscape, there is still a well-known deficit in the knowledge about the abiotic (geologic) factors that form the landscape, its age, and its origin. The ultimate objective of our study is to interpret the landscape of the Urdaibai Biosphere Reserve, located 40 km northeast of Bilbao (figure 4.1), from a geological perspective and to examine its evolution during the Quaternary period.

A *biosphere reserve* is a geographical place that is representative of the planet's different habitats, including both land and marine ecosystems. An

---

* This study is part of the Research Project, "Paleoenvironmental Evolution of the Urdaibai Region in the Pleistocene Based on Karstic Fills: Preliminary Study" (05/025), underwritten by the Unesco Chair of the Basque Government. Special thanks to the Gernika speleological group ADES and to Christophe Falgueres and his group for the dating provided. This is the third contribution to the Geo-Q Center's research activities.

ecosystem is defined as a natural system formed by a combination of live organisms (biocenosis) and the physical environment where they interact (landscape and geology). Considering that the Urdaibai Biosphere Reserve is located in a mountainous region and is essentially carbonate, the Quaternary evolution of the abiotic landscape is, at least in part (or essentially), collected in the karstic systems.

Figure 4.1. Geographical location of the Urdaibai Biosphere Reserve area.

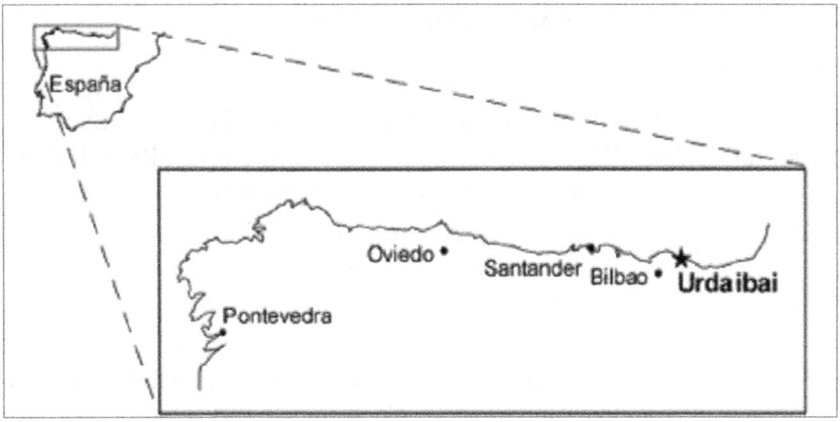

Completion of this project has had a dual focus: a broad examination of the Urdaibai Biosphere Reserve area, looking at both the physical relief and cavities, and a study of several representative examples. Of all of the 233 caves described by the ADES Speleological Group in the calcareous zone that represents 10 percent of the reserve's total area, 3 have been studied in relation to the spot height of the entrance, inactivity, detrital/chemical sediment, and proximality/ distality to the coast.

In addition to the geological, archaeological, and anthropological values that the karst landforms hold in their interior, their sociocultural and economic interest is also worth noting. Ultimately, the caves allow us to travel several million years back in time, inviting us to imagine how they were formed, the pervading climate, the type of landscape that existed, and the role that many of them could have had for various human cultures. In addition, the caves offer a very different environment than the one we live in, one distinguished by absolute darkness, the sounds of falling droplets over silence, and the formation of mineral structures that, when illuminated, transform into true works of art. Successful tourism at caves that are open to the public demonstrates that they continue to hold interest for society. However, the information we have about their origin, formation age, and the evolution of the karstic system itself is very limited.

Karst Landscape and Caves

Karst topography is an area with inherent hydrological characteristics and geomorphological features generated as a result of the combination of a soluble geological layer (limestone, gypsum, etc.), and a well-developed secondary porosity (galleries, sinks, caves, etc.).

The term *karst* comes from a region in Slovenia characterized by the presence of limestone and soluble bedrocks (not always carbonates) where water drainage occurs through extremely ramiform subterranean cavities. The interior water flow through the karstic system can be identified by tracers (colorants, isotopes, floating items, etc.) that are poured into the subterranean watercourse and then observed upon their exit. The geomorphology of karst areas is very unusual in that it is dominated by dissolution structures such as dolines, poljes, sinks, lapies, tunnels, and caves, all of which are formed by the action of the water.

Urdaibai is distinguished by the development of enormous dolines and caves. The caves are natural cavities formed in a rocky body. They are connected to the surface and present a variety of dimensions and unequal development, from a few meters up to several kilometers. Dissolution is the primary mechanism in the formation of the cavities. There are several factors that intervene in and control said process. The presence of soil with plant matter and the mechanical characteristics of the affected rock are among the most important.

When water passes through organic soil, it incorporates the carbon dioxide (CO_2) from the decomposition of the organic material, forming a mild acid called carbonic acid:

$$H_2O \text{ (water)} + CO_2 \text{ (carbon dioxide)} \rightarrow H_2CO_3 \text{ (carbonic acid)}$$

When the water, now acidified after passing through the organic soil, arrives at the weak surface, comprised of fractures or planes in the limestone, the carbonic acid reacts with the limestone, increasing the size of the fracture or initial discontinuity, and thus begins the formation of the cavities:

$$H_2CO_3 \text{ (carbonic acid)} + CaCO_3 \text{ (calcite)} \rightarrow Ca^{++} \text{ (calcium ion)} + 2\ HCO^{3-} \text{ (bicarbonate)}$$

Limestone that crops out in Urdaibai is primarily from the Jurassic and Cretaceous periods, and presents lithologies primarily formed by calcite, a carbonate mineral composed of calcium carbonate ($CaCO_3$). Karstic cavities provide a favorable sedimentary medium for depositing various types of sediments, including detrital, chemical, and organic.

The detrital sediments (figure 4.2a) are composed of particles resulting from the erosion, transport, and sedimentation of other rocks or pre-existing deposits, or by soil dismantling. In the cave deposits, detritial sediments may be transported to the karstic system from the cave's exterior (allochthons), or they may be the product of an alteration in the rock that hosts the karstic system (autochthons).

The various precipitate minerals that give the dark caves light and color are simply secondary precipitates (chemical sediments, or speleothems) that are formed from saturated water droplets or flow and are found primarily in the vadose zone (unsaturated) of the karstic system. The carbonate needed for precipitation of speleothems comes from the dissolution of the limestone itself during the formation of the cavities. The acidic water, loaded with CO_2, dissolves the limestone while it infiltrates the rocky massif. When the water arrives at the roofs of the cavities, the pressure conditions vary and the excess of CO_2 is released, depositing a small amount of calcium carbonate in the mineral form of calcite or aragonite onto the substratum. The slower the deposit process, the more perfectly the crystal is formed (figure 4.2b).

Figure 4.2. (a) Example of detrital sedimentary fills in the Aretxalde Cave (Urdaibai). (b) Speleothems from the interior of a cavity in Urdaibai. (c) Organic fills formed by osseous remains. Source: Images provided by the ADES Speleological Society.

In the various cave-fill deposits, the organic component is directly related to the presence of bat or bird colonies (especially in coastal areas) or animal or human occupancy of the caves. Thus, there is a greater organic accumulation in the entrance zone of the caves or shelters that is less commonly found toward the interior of the caves (unless it is redeposited or in a secondary position). This organic contribution is provided in the form of accumulations of excrement or guano (bats and birds); accumulations of bones from bezoars of large nocturnal birds (e.g., Eurasian Eagle-Owl); accumulations of food remains from carnivores (animals and hominids); or the skeletons of animals and hominids that lived in or occupied the cavities (figure 4.2c).

In this regard, it is important to point out the extensive occupation of certain cavities in the Urdaibai environment by hominids in prehistoric periods, as shown in archaeological sites such as Antoliña and Santimamiñe. In the cave-fill sediment of those sites, there is an abundance of food remains such as shells and large herbivore bones along with tools for hunting or food handling.

It is also common to find in the cave sediments accumulations of pollen remains from the plants that currently and formerly surrounded the karstic environment, perhaps carried by the wind or water. Pollen analysis is a very useful tool for identifying the ecosystems and thus the climate that predominated in the past.

Geomorphology and the Urdaibai Landscape

The geomorphological features that dominate in the Urdaibai Biosphere Reserve are of a karstic nature as opposed to coastal (platform) or fluvial (valley and terrace erosion). On the coastline, relict structures predominate, such as elevated abrasion platforms ("rasas") that are indicators of tectono-eustatic restructuration that occurred throughout the Cenozoic era, along with estuaries, or river valleys flooded by the sea and converted into rias that have an estuary circulation after the last climate change (11,500 years ago).

Karstic modeling primarily affects Urgonian limestone (Aptian-Albian). It is characterized by large, irregular dolines (Basondo, Ondobakozulo, Aldekorta, Iruskieta, Matxibarre, etc.) formed from an erosion surface and leaves relict reliefs that are pinnacle, conical, or even hemispherical in shape; more or less isolated; and of a similar spot height. It could be defined as a "polygonal karst" typical of humid temperate and tropical regions (Williams 1971, 1971). Seen from the air, it resembles an egg box–like topography, generated by dissolution from specific surfaces that, in the case of Urdaibai, are found at 400 m to the south and 300 m along the coast ("fundamental surface"). These surfaces, possibly erosive in nature and therefore able to generate an exokarst model because of descending dissolution, extend from the current coast following the river valley 12 km inland. The Bermeo and Mundaka horizontal platform and perhaps also the Asnarre promontory, located at altitudes around 50–60 m and 20 m, respectively, have emerged marine or (paleo) tidal platform erosion surfaces and their spot height concurs with the platforms described as Uribe-Kosta and Castro Uriales.

The pinnacle landform that can be seen in Urdaibai corresponds to the evolution of a modeling, perhaps pre-Quaternary (decidedly upper pre-Paleolithic - 40 ka), that was modified by the introduction of the Quaternary river network. In the interior of the Iberian Peninsula, it is common to find karst platforms above altitudes of 1000 m. However, as we move closer to the coastline, the carbonate platforms begin to decrease in spot height and a leveling of altitude can be detected that corresponds to specific spot heights and occurs at levels close to 500 m, a bit above 300 m, and between 100–150 m (Mediterranean coast). The fundamental karst platform or surface of Ereñozar, without overlooking the tectonic, corresponds with the intermediate spot height.

On a lesser scale, the exokarst is characterized by lapiés, which are essentially penetrative and clearly observable in Antoliña or in the vicinity of Atxerre. Despite being currently almost entirely covered by the Cantabrian Holm Oak forest, it is possible to distinguish lapiés generated in favor of planar heterogeneities, such as joints and stratification surfaces, and channelized shapes parallel to the maximum gradeability line that are as a result of the direct dissolution of rainwater (not related to rock heterogeneities).

Karstic Cavities in Urdaibai

Pursuant to the studies performed, we can conclude that the karstic cavities in Urdaibai demonstrate a control that is strongly structural (e.g., Laga or Asnarre paleokarst) or stratigraphic (e.g., Aretxaga and Antoliña).

In Ereñozarre, considering the different spot heights of the cave entrances, the karstic cavities are grouped into set spot heights that are not random: at altitudes of 40 m, 60 m, around 180 m, and between 270 and 300 m (heights that coincide with the "fundamental surfaces"). The mouths of the caves are just the intersection of the karstic system with the slope, in this case, created by the incision of the doline and/or the river valley. And given that the cavities studied are of a phreatic nature, developed in the saturated zone, they may indicate the position of the different base levels along the engagement of the river valley. For that reason they have been identified as the fossil caves with the highest spot heights (Antoliña, Aretxaga) and the lowest (Laga, 20 m).

Cavity Fills

All of the cavities studied show one or several fill sequences comprised first of a detrital sedimentary fill and then accompanied by a phase of chemical precipitate (speleothems). The detrital sediments are, for the most part, allogenic, not related to the karst massif and, therefore, come from the exterior (allochthons). The abundance of very rounded clasts is important, as it provides evidence of water transport and siliciclastic (sandstone, white quartzite, and lutites corresponding to the lithostratigraphic units of the Supraurgonian), as well as igneous (ophitic) environments. Particles formed in soils, however, are also abundant (e.g., limonites). In the Antoliña cave, river sediments have not been observed. In contrast, at the cave entrance, allochthon sediments transported by gravity as well as anthropogenic and organic (phosphate) sediments predominate.

Asnarre Paleokarsts

The Asnarre Promontory is located in the northeast part of the Urdaibai Biosphere Reserve, at the end of the sand on Laga beach and at the foot of the Ogoño cape (figure 4.3). It is formed by the Otoio Cliff Formation (Agirrezabala 1996), a carbonated unit of limestone with rudists (caprotinid) (figure 4.4) from the upper Aptian-lower Albian period. The promontory features a north–south position and subvertical dip and is limited at the base by siliciclastic units and at the roof by marl units. The limestone in the formation reaches a stratigraphic thickness of 55 m. (Agirrezabala 1996).

Figure 4.3. (a) Aerial photo of Laga beach. (b) Side view of the Asnarre Promontory and Cape Ogoño from Laga beach.

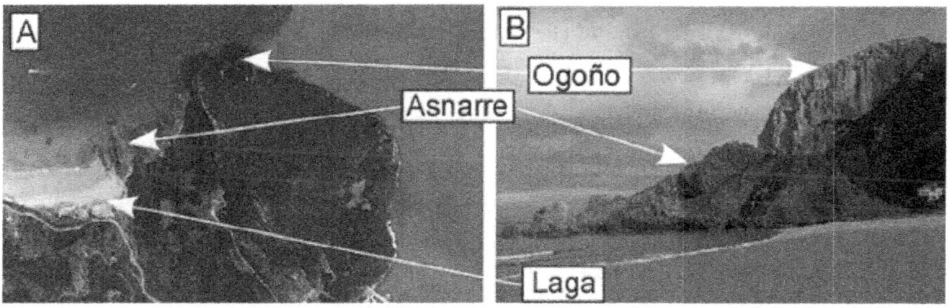

Figure 4.4. Carbonated rock in the Otoio Cliff Formation that comprises the Asnarre Promontory. Ru: Cross-section of caprotinid rudists (extinct bivalve mollusks); Ch: Chondrodonta sp. shells (extinct bivalve mollusks); Mt: Micritic matrix of the rock.

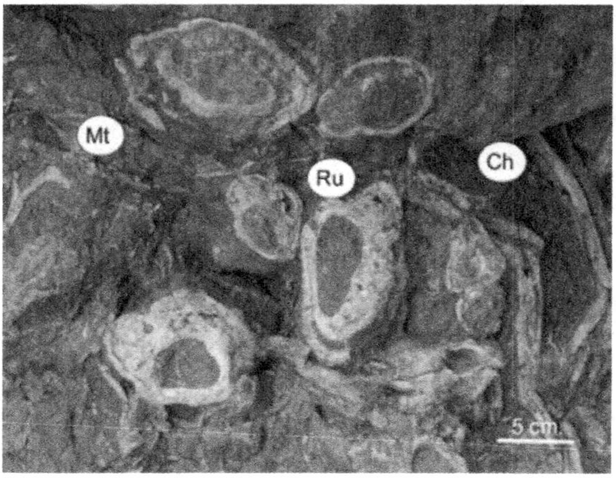

Differential erosion and gravitational collapse have allowed the paleokarst created in the interior of the limestone massif, which is now found to be inactive, to be exposed along an approximately 50-m escarpment (figure 4.5A and B). The karstic dissolution of Asnarre continues to be a clearly structural directrix. The strong diaclase created along directions N60°E (56/150) and N70°E (50/340), perpendicular to the stratification (85/90), limited development of water infiltration lines and gallery formation in the areas of diaclase intersection.

Along the escarpment, two levels of sub-horizontal galleries can be distinguished that have metric dimensions, are phreatic, and are positioned east–west (parallel to the coastline). These galleries have a minimum horizontal development of 50 m and are located at 5 m and 20 m above current sea level (figure 4.5a and b). The galleries at the highest level (20 m) are the ones that most crop out. These phreatic levels are separated vertically by vadose flows generated in favor of the dissolution of diaclases.

Figure 4.5. Asnarre karst: (a) and (b) phreatic karst galleries; (c) and (d) fissure fills.

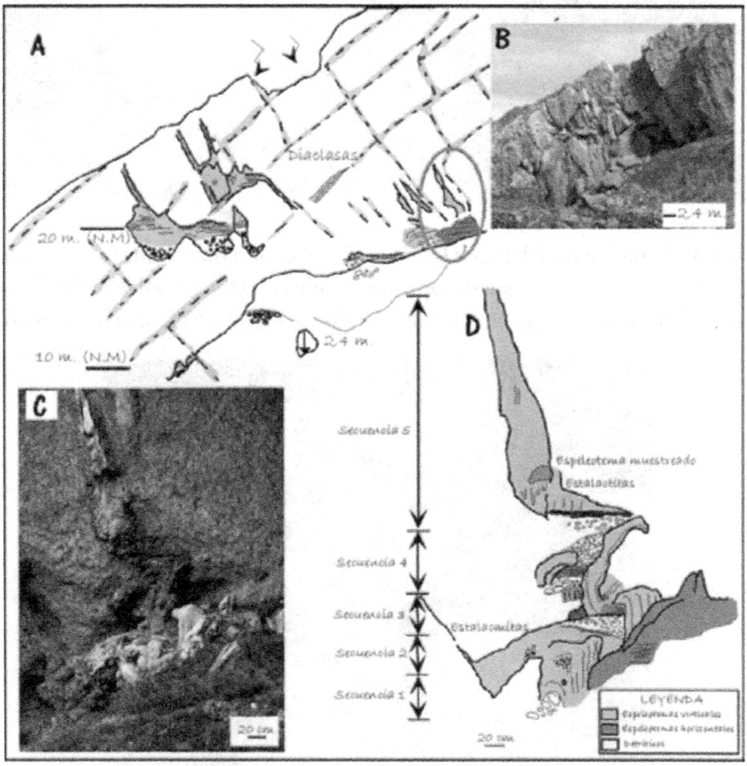

Asnarre's Paleokarst Fills

The sedimentary fill of the paleokarst is dual in nature: detrital (siliciclastic) and chemical (calcite). In the more developed cavities, the detrital fill is confined to the basal parts, while the chemical precipitate fills the cavities and fissures. This sedimentary sequence may be repeated up to five times in the vertical, demonstrating a polyphasic fill of the cavities (figure 4.5c and d).

Detrital sediments are comprised of polymictic orthoconglomerates with a sand matrix and some angular fragments of limestone produced by the wall's gravitational fall. The levels appear with a sub-horizontal disposition and are cemented by calcite. The average diameter of the clasts is 20 cm. (although they vary between 3 cm and 50 cm). Very rounded cylindrical formations predominate and are siliciclastic (primarily quartzarenites, limonites, quartzites, and ophites) (figure 4.6a). The allochthonous nature of the detrital sediments, their presence exclusively in phreatic galleries, and the absence of marine fossils in the sand matrix that holds them imply that their deposit is of a fluvial-karstic origin (allogenic karst) (Lauritzen 1990).

Figure 4.6. Sedimentary fills of the Asnarre paleokarst: (a) allochthonous detrital sediments; (b) column; (c) detrital clasts fossilized by speleothem that is laminar; (d) cross-section of stalagmites.

Chemical sediments mark the cessation of the detrital traction supports. Various types of speleothems are distinguishable (figure 4.6b), such as flowstones, stalagmites, stalactites, and columns, although the detrital sediments are always fossilized by flowstones that cement the underlying sediments, forming a laminate deposit (figure 4.6c), sometimes with internal erosive scars that adapt to the clasts. The formation of flowstones is followed by drip formations such as stalactites and stalagmites (figure 4.6d).

Uranium/Thorium (U/Th) dating performed on the Asnarre speleothems at the Prehistory Department, Muséum National D'Histoire Naturelle in Paris, France, reflect an average formation age for the vadose phase of the first sequence (figure 4.6) to be around 300 ka (table 4.1).

Table 4.1. First vadose sequence speleothem dating of the Asnarre Promontory karst cavity fills.

Laga	Mineral	U ppm	U^{234}/U^{238}	Th^{230}/Th^{232}	Th^{230}/U^{234}	Age ka
LAG0701	Calcite	0.07	1.222 ± 0.032	53	0.983 ± 0.032	292 +65 / −40
LAG0702	Calcite	0.08	1.340 ± 0.044	29	1.037 ± 0.044	339 +166 / −66

Conclusions

The karstic cavities of the Asnarre Promontory present an elongation parallel to the coast and limited development of the phreatic sub-horizontal galleries toward the continent. This could be evidence of the influence of a mix of freshwater and saltwater in the formation of the cavities, in the same manner described by Alfred Bögli (1980) as well as John E. Mylroie and James L. Carew (1990). This mix of water could have occurred during the elevation of the Asnarre marine erosion platform (rasa).

The phreatic cavities show a repetitive-fill sedimentary sequence that marks a decreasing energy trend from siliciclastic fluvial gravel related to periods of high transportability (rain and strong water flow), to the formation of flowstone speleothems created by laminate water flows, and ending with the filling of the cavities with drip speleothems corresponding to the low (relative) water flow periods and/or decrease in the active phreatic level (base level). Meanwhile, the vadose cavities present solely laminate (flow) or drip (straws and stalactites) speleothem formations. Based on comparison with the Pindal Cave model (Jiménez-Sánchez et al. 2006), the Asnarre paleokarst may correlate with the elevated marine platform that bears the same name and that today is located at 50 m to 60 m along the Asturian coast. The U/Th dating obtained for the speleothems indicates that the elevation of the Asnarre Promontory's tidal platform and the beginning of the dissolution of the karstic conduits therein took place before the middle Pleistocene (±300 ka).

References

Agirrezabala, Luis Miguel. 1996. "El Aptiense-Albiense del Anticlinorio Nor-Vizcaíno entre Gernika y Azpeitia." PhD diss, University of the Basque Country.

Bögli, Alfred. 1980. *Karst Hydrology and Physical Speleology*. Berlin: Springer-Verlag.

Ford, Derek, and Paul Williams. 2007. *Karst Hydrogeology and Geomorphology*. Chichester, UK and Hoboken, NJ: Wiley & Sons.

Gillieson, David S. 1996. *Caves: Processes, Development and Management*. Oxford and Cambridge, MA: Blackwell.

Jiménez-Sánchez, Monserrat, J.L. Bischoff, H. Stoll, and A. Aramburu. 2006. "A Geochronological Approach for Cave Evaluation in the Cantabrian Coast (Pindal Cave, NW Spain)." *Zur für Geomorphologie* 147: 129–41.

Lauritzen, Stein-Erik. 1990. "Autogenic and Allogenic Denudation in Carbonate Karst by the Multiple Basin Method: An Example from Svartisen, North Norway." *Earth Surface Processes and Landforms* 15, no. 2: 157–67.

Mylroie, John E., and James L. Carew. 1990. "The Flank Margin Model for Dissolution Cave Development in Carbonate Platforms." *Earth Surface Processes and Landforms* 15, no. 5: 413–24.

Williams, Paul W. 1971. "Illustrating Morphometric Analysis of Karst with Examples from New Guinea." *Zeitschrift fur Geomorphologie* 15, no. 1: 40–61.

———. 1972. "Morphometric Analysis of Polygonal Karst in New Guinea." *Geological Society of America Bulletin* 7, no. 3: 761–96.

5

Recent Environmental Transformation of the Bilbao Estuary: Natural and Anthropogenic Processes

ALEJANDRO CEARRETA, Mª JESÚS IRABIEN
University of the Basque Country (UPV/EHU)

EDUARDO LEORRI,
East Carolina University

The Bilbao Estuary is located on the northern coast of Spain (43°23'–43°14'N, 3°07'–2°55'W; figure 5.1). Originally Spain's most extensive estuarine area (Hazera 1968), the modern estuary is 15 km long and an average of 100 m wide, and its channel depth ranges from 2 m in the upper estuary to 9 m at the mouth. It is formed by the tidal part of the Nervion River, although four other rivers (Kadagua, Asua, Galindo, and Gobelas) discharge into the main course (figure 5.1). Tides range between 4.6 m (spring tides) to 1.2 m (neap tides). The estuary is partially mixed and the annual average freshwater inflow is 25 $m^3 s^{-1}$ (García-Barcina, González-Oreja, and de la Sota 2006), compared with a sea water input of approximately 230 $m^3 s^{-1}$ (Urrutia 1986). The metropolitan area of Bilbao and its surroundings are located along the axis of the Bilbao Estuary with more than 1 million inhabitants and 2,700 industrial facilities (Borja et al. 2002; García-Barcina, Oteiza, and de la Sota 2002).

Historically, the natural features of the Bilbao Estuary have been dramatically modified by urban, industrial, and port developments. The exploitation of abundant local iron ore led to the early industrial development of Bilbao in the mid-nineteenth century. The original estuary was rapidly reduced in size through land reclamation to form a tidal channel (completed by 1885) as a navigable watercourse from the city to the open sea (Junta de Obras del Puerto de Bilbao 1910). The first iron and steel industry was opened on the middle estuary over reclaimed marshes as early as 1854 (García-Merino1987). Since this time, the Bilbao Estuary has received waste from many sources

* This work constitutes a summary of previously published papers by the authors on the Bilbao Estuary and has been partially funded by the IT365-10 project (Harea-Coastal Geology Group, Basque Government). It represents Contribution #2 of the Geo-Q Research Unit (Joaquín Gómez de Llarena Laboratory).

Figure 5.1. Geographical location of the Bilbao Estuary, position of the boreholes, and localities referred to in the chapter. Black dots indicate boreholes studied in Leorri and Cearreta (2004). White dots indicate boreholes previously studied in Cearreta (1998). Gray dots indicate other boreholes drilled during the geological study for the Bilbao Metropolitan Subway. Dashed lines represent the original extent of Holocene estuarine domains. Mid-gray shaded surfaces are urban areas (modified from Leorri and Cearreta, 2004).

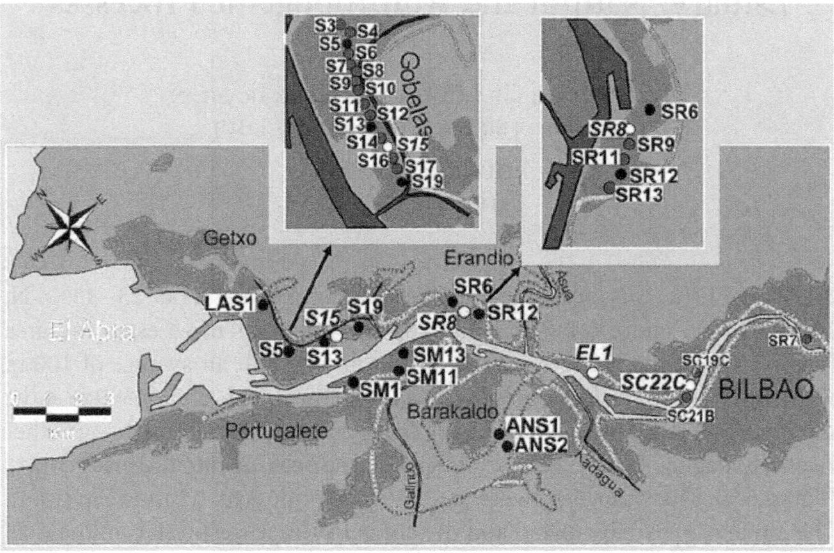

(mineral sluicing, industrial waste, and urban effluents) that have significantly degraded its environmental quality. As a consequence, its water and sediments have had extremely low concentrations of dissolved oxygen, and a high content of organic matter and heavy metals (Irabien 1993; Sáiz-Salinas, Francés-Zubillaga, and Imaz-Eizaguirre 1996). However, there has been a significant decrease in the flux of organic matter and heavy metal contaminants in recent decades due to the implementation of environmental protection policies, an improvement in waste-treatment systems, and the closure of some major factories (Gorostiaga and Díez 1996). A sewerage scheme was implemented in 1984 comprising a combined sewer network that conveys sewage to a central wastewater treatment plant (WWTP). Since 2001–2002, the plant has had a biological treatment capacity (García-Barcina, González-Oreja, and de la Sota 2006).

Today, the Bilbao Estuary is a largely artificial system that bears little resemblance to the original estuary. It has been calculated that the total amount of the original estuarine surface lost through human activity is approximately 1,000 ha (Rivas and Cendrero 1991).

Formation of the Original Bilbao Estuary

Numerous geotechnical studies carried out during the period 1879–1995 by the Bilbao Port Authority (Puertos del Estado 1997) show that the average thickness of the Quaternary sequence in the Bilbao Estuary varies greatly from 10 m in the upper, 20 m in the middle, and 30 m in the lower-estuarine areas, with basal sandy gravels of possible fluviatile origin along the central axis of the original pre-industrial estuary. Furthermore, within the same estuarine area, basement can be found at extremely variable depths even among very closely drilled boreholes. Alejandro Cearreta (1998) and Eduardo Leorri and Cearreta (2004) reconstruct the palaeoenvironmental changes in this marginal marine environment since its formation following the postglacial rise of sea level. They analyze the microfossil (benthic foraminifera) content of numerous boreholes taken in reclaimed areas along the estuary (figure 5.1). Integration of those results with similar data from previously studied boreholes allows us to reconstruct the environmental development of the Bilbao Estuary during the Holocene. This environmental development has been organized into different systems tracts following a sequence stratigraphic interpretation (Allen and Posamentier 1993). Radiocarbon dating help to locate in time the different palaeoenvironments and depositional episodes identified in the boreholes.

Figure 5.2 shows the sequence stratigraphic interpretation of the Bilbao Estuary fill based on microfossil assemblages present in the boreholes drilled on the right bank of the estuary (figure 5.1). The sequence boundary (SB) is defined by the location of unconsolidated Quaternary gravels and sands over Cretaceous basement. This boundary is overlain by a deposit of fluviatile gravel and coarse sand that corresponds to the basal interval in the upper- and middle-estuarine areas. These fluvial deposits, barren of foraminifera, are interpreted as the low-stand systems tract (LST). Due to the lack of carbonaceous material, this LST has not been radiocarbon dated, but its stratigraphic position underlying lower-Holocene estuarine deposits suggests a late-Glacial and lowermost-Holocene age. In the lower-estuarine area these coarse sediments were reworked during the Holocene marine transgression and, consequently, were included in the following systems tract.

Accumulation of estuarine sediments containing abundant foraminifera over fluviatile coarse deposits clearly suggests a marine transgression within the former fluvial valley. Different intervals were deposited during this transgressive episode, all of them exhibiting increasing numbers of foraminifera and, consequently, indicating an upward increase in open-marine influence as transgression progressed with time. These intervals comprise the transgressive systems tract (TST) bounded by distinct stratigraphic surfaces. A transgressive surface (TS) can be defined at the base of these estuarine depos-

Figure 5.2. Sequence-stratigraphic interpretation of the Bilbao Estuary Holocene deposits. Vertical lines indicate borehole locations on the right bank of the estuary drilled during the geological study for the Bilbao Metropolitan Subway. LST: low-stand systems tract; TST: transgressive systems tract; HST: high-stand systems tract. Depth with reference to local ordnance datum (modified from Leorri and Cearreta 2004).

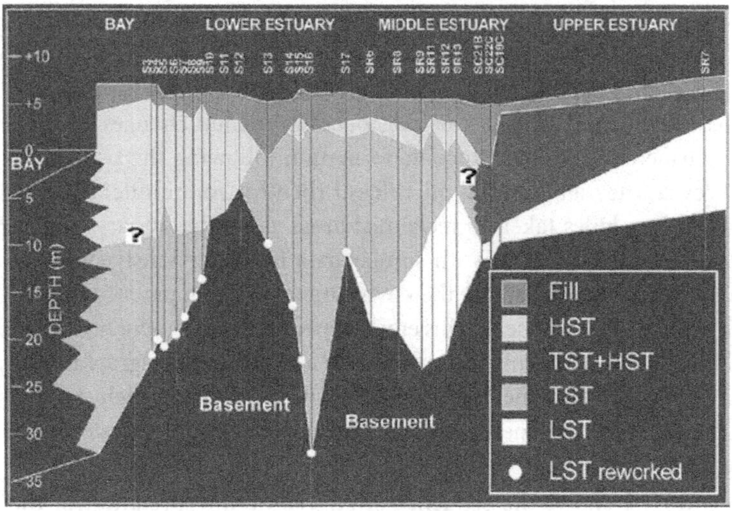

its separating them from the coarse LST material. As the transgression continued, the estuarine materials were overlain by sediments from the inner shelf (containing dominant marine foraminifera) that increased near-marine conditions within the estuary. Radiocarbon dates obtained in this TST range from 8520 to 2685 cal years BP.

Different curves from the Bay of Biscay show that sea level reached approximately its present position at about 3000 cal years BP (Pirazzoli 1991). The sedimentary response to this stabilization of sea level was the turnaround from eustatic transgression to relative regression, with deposition of estuarine materials similar to those of the underlying TST, but containing brackish and shallower foraminiferal indicators. The uppermost intervals in the boreholes comprise this high-stand systems tract (HST). The surface between the HST and the TST represents the maximum flooding surface (MFS). Radiocarbon dates obtained from HST materials were younger than 2810 cal years BP.

In summary, sediments that comprise the original estuarine infill range from fluviatile through brackish estuarine to near-marine and even supratidal materials. Analysis of sedimentological and micropalaeontological data indicates that, in general, boreholes exhibit fining-upward sequences ranging from gravelly materials at the base, more sandy sediments in the middle, and more muddy materials on the top. At the same time, basal coarse materials are barren of foraminifera; middle-sandy sediments show abundant, diverse, and near-marine foraminiferal assemblages; and the upper-muddy deposits contain abundant, low-diversity, and brackish foraminiferal assemblages. Micropalaeontological interpretation together with radiocarbon dates suggest that basal coarse materials are late-Glacial and lowermost-Holocene fluvial deposits that are followed upward by lower- and middle-Holocene transgressive materials, with upper-Holocene regressive sediments on top. The sedimentary and microfaunal features of this general sequence are variable depending on its palaeogeographic setting: increasing muddy sediments and absence of open-marine elements are characteristic of the upper estuary, whereas increasing sandy sediments and reworking of the basal fluviatile materials are typical of the lower estuary.

Industrial Transformation of the Bilbao Estuary

Heavy metals and organic contaminants discharged into the coastal environments may admix or interact with fine-suspended sediments and accumulation of contaminated sediments in salt marshes, and inter- and sub-tidal mudflats can cause these contaminants to be trapped within estuaries. Heavy metals are not necessarily fixed permanently in the sediment, however, and may be released to the water column by physical reworking (e.g., dredging) and early-diagenetic processes, even when the untreated discharges have stopped (Beno and Gibbs 1990). Thus these contaminants may continue to have significant environmental impacts even after discharge limits and cleanup procedures have been implemented.

Records of environmental change are often preserved in sediment from sub-tidal and inter-tidal mudflats and salt marshes, the study of which allows one to assess the scale of pollutant inputs from past development (Valette-Silver 1993; Croudace and Cundy 1995; Cundy et al. 1997). Such studies aid evaluation of effluent cleanup and environmental remediation measures. By concurrently analyzing microfaunal remains, the historical impacts of pollutants on estuarine ecosystems can be assessed. While selective preservation of organisms limits the extent to which ecosystem-wide changes can be assessed, pollution-induced changes have successfully been examined using organisms that have good preservation potential, such as foraminifera (Alve 1991; Sharifi, Croudace, and Austin 1991).

Cearreta et al. (2000; 2002) used an integrated chemical–microfaunal approach to examine the recent history of pollution in the Bilbao Estuary, offering geochemical data and benthic foraminifera from sediment cores (up to 10 m long) collected from the highly polluted intertidal flats. Determinations of the deposition rates of ^{210}Pb and ^{137}Cs were also undertaken to provide a chronology for pollutant inputs and ecological changes in the Bilbao Estuary.

Within the sedimentary record matching the industrial period (containing high concentrations of metals), it was possible to differentiate two zones (figure 5.3): an older industrial zone characterized by the coexistence of an abundant foraminiferal assemblage with high levels of metals, and a younger industrial zone with extremely high levels of metals and barren of estuarine foraminifera.

Figure 5.3. Environmental interpretation of the Bilbao Estuary industrial cores based on the combination of microfaunal and geochemical data. Depth with reference to local ordnance datum (modified from Cearreta et al. 2002).

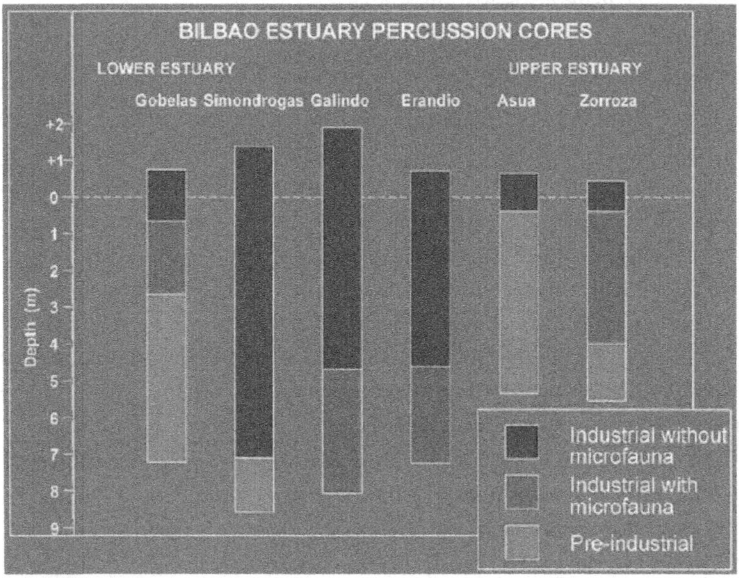

In the percussion cores in which the pre-industrial zone is followed upward by the older industrial zone, their respective foraminiferal assemblages are similar in terms of species dominance. However, the older industrial assemblage shows a reduced number of individuals and species than the pre-industrial assemblage. The relatively erratic distribution of radionuclides reported by Cearreta et al. (2000) makes it impossible to date these materials and, consequently, to determine if both zones are continuous in time or if they are separated by time lags provoked by dredging carried out along the estuary since the nineteenth century. This older industrial zone clearly indicates that at one pint a normal foraminiferal assemblage could live and reproduce in a moderately polluted estuarine environment, probably during initial industrial development of the area (1850–1950), as can be deduced from written historical records (García-Merino 1987).

On the other hand, the younger industrial zone is present as a mantle covering the uppermost part of every core, although the thickness of this heavily polluted zone varies greatly along the estuary. The highest volumes of the "polluted mantle" are found in the middle-estuary area where dredging has been historically more intense (figure 5.3). This zone represents later-industrial development in the Bilbao Estuary (1950–2000) with major urban and industrial discharges that resulted in minimum oxygen levels and extremely high organic matter and metal concentrations. As a consequence, complete defaunation followed in the estuary under degraded environmental conditions that persisted during many years.

The high concentrations of heavy metals and arsenic found in the studied cores reflect the substantial degradation of the Bilbao Estuary. However, the data obtained exhibit considerable spatial variability. The lowest levels of metal enrichment were found in the upper estuary, while the highest values appear in the middle part. This heterogeneous distribution reflects the proximity-to-point sources of pollution, which discharge directly into the estuary or into the lower reaches of the four main tributaries. Similarly, abundant foraminiferal assemblages coexisted with very high metal concentrations in the older industrial zone of the middle-estuarine area, whereas they had disappeared in the younger industrial zone of the upper-estuarine area under lower levels of metal pollution. This supports the idea that oxygen limitation is the key factor in explaining the complete defaunation of the Bilbao Estuary as reflected in the newer industrial zone, although high levels of pollutants also contributed to the severe historical degradation of this environment (Sáiz-Salinas, Francés-Zubillaga, and Imaz-Eizaguirre 1996; Cearreta et al. 2000).

The variations of metal concentrations by depth within the studied cores may be a consequence of changing anthropogenic inputs, early diagenetic reactions, and physical reworking of sediments. Unfortunately, we lack detailed historical archives that document temporal trends in anthropogenic fluxes into the Bilbao Estuary. However, the vertical distribution of metals in the studied sediment cores suggests that, historically, concentrations of zinc, lead, and copper (and probably arsenic) started to increase earlier in the estuary, followed by chromium and nickel.

Modern Environmental Regeneration of the Bilbao Estuary

As noted, there has been a significant decrease in the flux of organic matter and heavy metal contaminants in recent years. Furthermore, estuarine regeneration is currently being undertaken as part of a strategic revitalization plan that includes: a) preventive measures, such as a sewerage scheme involving a network of interceptor and collector sewers with major wastewater treatment plants; and, more recently, b) remedial action involving dredging of heavily contaminated surficial sediments, construction of safe and controlled disposal sites for dredged material, and, possibly, the implementation of sediment decontamination treatments. Despite these improvements, however, the unremoved contaminated sediments from the inter-tidal areas may act as a long-term source of heavy metals to the aquatic environment through sediment mechanical reworking (e.g., dredging and shipping) and oxidation of anoxic sediments.

The European Water Framework Directive (WFD; Directive 200/60/EC 2000), considered the most significant piece of legislation regarding water quality (Andersen, Conley, and Hedal 2004), provides national and local authorities with a legislative basis for maintaining and recovering water quality to achieve a good ecological and chemical status for all surface waters (Borja 2005). Although there is a lack of reference to sediments in the WFD, it contains potential implications for managing these materials. Under the WFD the purpose of analyzing the levels of priority substances in sediment might be to monitor the progressive reduction in the contamination of priority substances and phasing out of priority hazardous substances as well as to demonstrate conditions of "no deterioration" in sediment quality (Stronkhorst et al. 2004).

The combination of living organisms and geochemical data provides a multi-proxy approach that can show any improvement of the sediment quality while also allowing assessment of this improvement in relation to the benthic biota. Sediment quality studies have usually relied on the use of different indices, yet the low values of abundance and diversity of biota in the Bilbao Estuary make any statistical approach irrelevant. Furthermore, biotic indices such as the AMBI (Borja, Franco, and Pérez 2000) do not indicate the almost abiotic nature of a heavily polluted area such as this and, therefore, raise some doubt regarding the applicability and/or relability of these indices (Quintino, Elliott, and Rodrigues 2006).

The only published monitoring program in the Bilbao Estuary that analyzes data since at least the 1990s deals with water quality (García-Barcina, González-Oreja, and de la Sota 2006). However, low residence times, low freshwater flows compared with volumes of sea water, current velocities, and tidal influence might give values of water parameters that are not representative of the environmental status and are subject to sudden changes (Bald et al. 2005). In contrast, sediment geochemistry and biota integrate environmental conditions and their variations occurred through time in a very effective way (Salas et al. 2006).

Leorri et al. (2008) reflect the results of the regular monitoring of foraminiferal contents and heavy metal concentrations in surficial sediments from the highly polluted inter-tidal flats of the Bilbao Estuary from 1997 to 2006, when significant reductions in contaminant sources and improvements in waste treatment were implemented. Current conditions were assessed using reference pre-industrial conditions obtained from cores recovered previously from the same estuary. In discussing spatial distributions and temporal trends in contamination, they offer a critical measure of "environmental health" for this estuarine area.

Monitoring the Bilbao Estuary started in October 1997. At that time, severe hypoxia from the lower estuary upward and anoxic conditions in the entire upper estuary dominated the bottom waters (Sáiz-Salinas 1997; Uriarte and Villate 2004). Sediments were completely anoxic up to the water interface (Sáiz-Salinas and González-Oreja 1998). In 1997, dissolved oxygen values were almost 50 percent in surface waters and less than 60 percent in bottom waters (García-Barcina, González-Oreja, and de la Sota 2006). The estuarine channel was heavily polluted, with organic matter values double those found in the nearby estuaries while labile compounds were four times higher (Cotano and Villate 2006). Dissolved oxygen seemed to be the main reason for this (Sáiz-Salinas 1997; González-Oreja and Sáiz-Salinas 1999). Not only did this adverse situation affect the estuary, but also the polluted waters eventually reached the sea and caused environmental damages up to 17 km away from the estuary (Díez, Santolaria, and Gorostiaga 2003). Furthermore, faecal coliform concentration exceeded guideline values 95 percent of the time during the bathing season at the beaches of the inner bay (García-Barcina, Oteiza, and de la Sota 2002).

In line with these results, initial sampling recorded no living foraminifera in the upper estuary with very low numbers elsewhere (figure 5.4). This distribution was related to the tidal transport of living individuals within the estuary (Cearreta et al. 2000). Heavy metal values were especially high for copper, lead, zinc, and the metalloid arsenic. The highly variable distribution of the different elements resulted from local sources of industrial pollution (Landajo et al. 2004). Despite the fact that these values were extremely high in surficial sediments, previous works document elevated levels of metals and arsenic in cored samples from the middle estuary, which are in general even higher than those reported for surficial materials (Cearreta et al. 2000, 2002).

Environmental conditions in the Bilbao Estuary began to improve with the economic recession and the closure of the mineral washeries (Gorostiaga and Díez 1996). This improvement has been more noticeable since the construction of the Galindo central WWTP. Physico-chemical treatment

Figure 5.4. Temporal evolution of foraminiferal standing crops in the spring sampling periods (1997–2009) along the Bilbao Estuary. Note the logarithmic scale. Reference conditions correspond to the mean ±1 S.D. values found in the Santoña estuary by Cearreta (1988) (modified from Leorri et al. 2008).

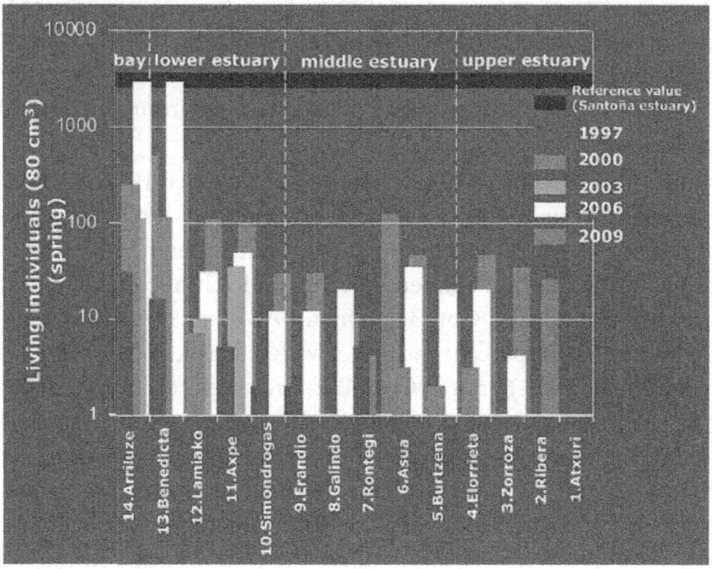

began in 1991 and ten years later biological treatment was initiated (Bald et al. 2005). This plant discharges a considerably higher water flux than the river flow into the Galindo River, especially during the summertime. Domestic sewage discharge is an important source of metal contamination (Ruiz and Sáiz-Salinas 2000), and the WWTP releases almost 50 percent of the inputs of chromium, lead, nickel, and cadmium, while for manganese it is as high as 77 percent (Landajo et al. 2004). Furthermore, during periods of intense rainfall, the Bilbao metropolitan area sewer system will mainly overflow through the WWTP outlet (García-Barcina, Oteiza, and de la Sota 2002).

Although the amount of pollutants discharged into the estuary has been decreasing gradually, this is not directly reflected in the sediments. In fact, José María García-Barcina, José Antonio González-Oreja, and Alejandro de la Sota (2006) show a significant reduction during 2000. However, no significant change was found in the period 1997–2000, either in the heavy metal content or in the foraminiferal assemblages. The remobilization of the sediments resulting from the intense dredging activities as part of the regeneration scheme reported by Mª Jesús Irabien et al. (2001) resuspended pollutants stored in the sediments during the last 150 years. Furthermore, intense oxygen depletion occurred during that year in the estuary (García-Barcina, González-Oreja, and de la Sota 2006). Dredged sediments in 2001 from the lower-estuarine channel and the inner bay were considered as probably toxic while the outer bay sediments showed intermediate levels of contamination (Casado-Martínez et al. 2006).

As a response to the biological treatment started in 2001, by 2003 water-dissolved oxygen had significantly increased, although the rise in the bottom waters was less pronounced (García-Barcina, González-Oreja, and de la Sota 2006). Additionally, heavy metal values and arsenic decreased up to 85 percent for some elements, this reduction was more intense for copper, chromium, lead, and zinc, and no samples exceed the ERMs (Effect Range-Median, proposed by Long et al. 1995) in the upper estuary, while only 25 percent of samples elsewhere failed more than two ERM values (figure 5.5). The substantial reduction of metal pollution observed from 2000 to 2003 is likely to be related to the implementation of biological treatment in 2001. The WWTP treats most of the urban sewage and the industrial effluents discharged into the estuary, reducing the average pollution loads. Since this plant came into operation, a significant improvement in water quality (García-Barcina, González-Oreja, and de la Sota 2006) and benthic recolonization (Borja et al. 2006) has been observed. Although this reduction is significant, we have to consider the potential environmental threat of the sediments as demonstrated by what happened in 2000. In fact, dredged sediments in 2003 from the lower estuary and the inner bay were still considered potentially dangerous (Casado-Martínez et al. 2006), reminding us that sediments below the surface still store huge amounts of contaminants. Bartolomé et al. (2006) report a similar improving trend between 1998 and 2003 for mercury. However, foraminiferal assemblages did not show significant changes. This is to be expected because colonization (faunal recovery process) can take from one to several years, and recently reoxygenated or severely contaminated sediments may delay this colonization (Alve 1999). Therefore, the closure of the major factories and the pollution abatement measures taken by local authorities seem to have resulted in a clear improvement in the geochemical quality of the accumulating sediments.

Finally, and contrary to expectations, in 2006 the number of samples classified in the medium-high priority category increased (figure 5.5). One plausible explanation for these significant perturbations in the overall reduction trend may be related to the remobilization of sediments by local dredging. Although the authorities have ruled out extensive dredging in sediment cleanup, focusing their efforts on reducing polluting loads and promoting natural regeneration, previous work seems to have led contaminated sediments to be removed and mixed upward into newly deposited cleaner materials. Activities such as dredging may resuspend fine materials into the water column and render pollutants to be bioavailable to biota (Nayar et al. 2003). In addition to the mechanical and biological effects, derived changes in redox

Figure 5.5. Proportion of surface samples classified in different categories based on the Effect Range-Median (ERM) quotients in the Bilbao Estuary through time (modified from Leorri et al. 2008).

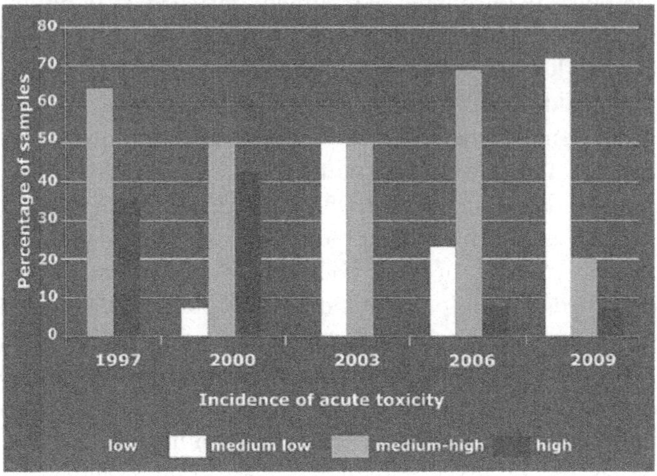

conditions may also affect metal mobility (Liang and Wong 2003). Unfortunately, there is no reliable information about the mechanisms of metals fixation in sediments from this estuary. Some authors (Irabien 1993; Sáiz-Salinas, Francés-Zubillaga, and Imaz-Eizaguirre 1996) suggest that precipitation of sulfides under anoxic conditions could be responsible, to some extent, for the scavenging of metals, but no geochemical evidence has been found (Cearreta et al. 2002). On the other hand, the inner bay area and lowermost estuarine samples show a significant quality improvement regarding foraminiferal analysis. In fact, an increase of two orders of magnitude was reached (figure 5.4), presenting similar values to those found in the Santoña Estuary (40 km to the west of Bilbao) with semi-natural conditions (Cearreta 1988). Furthermore, the assemblages resembled those recorded in pre-anthropogenic materials from Bilbao (Cearreta et al. 2002; Leorri and Cearreta 2004). This is probably a result of the tidal influence diluting the pollutants and increasing the levels of dissolved oxygen.

As a consequence of the severe degradation in the estuary during the last 150 years, its environmental regeneration is one of the major concerns of the local government. In 1993 the local water authority had envisaged a recovery of its native wildlife by the year 2002 (Consorcio de Aguas del Gran Bilbao 1993; Sáiz-Salinas 1997). Unfortunately, this has not been the case. In fact, water quality is not complying with the guideline values of the European Directive on Bathing Waters in the inner bay beaches (García-Barcina, Oteiza, and de la Sota 2002), and sediments might not achieve the quality status by the year 2015 as required by the European Water Framework Directive (Muxika, Borja, and Bald 2007).

The results obtained here imply a slow environmental recovery of the Bilbao Estuary in a similar way to that of the Mersey Estuary in Liverpool, England, where recovery has taken more than forty years since the worst pollution period. Indeed, here there are still problems in which the accumulation of heavy metals and persistent organics in animals such as fish are a potential human health problem (McCauley, DeGraeve, and Linton 2000). At this point, the central WWTP and the sediments that still contain pollutants should be a major concern as pollution sources. In order to assess the potential recovery of the Bilbao Estuary and provide sound data, long-term series based on heavy metals and foraminifera have been proven to be suitable. Recovery predictions are difficult to make, but establishing adequate reference conditions and continuous monitoring will provide the necessary temporal trend to anticipate future change and provide data to undertake remedial actions. As is the case with the Bilbao Estuary, frequent spills from the WWTP and, more important, sediment resuspension have to be remediated if water and sediment quality objectives are to be met.

Conclusions

The original structure of the Bilbao Estuary sedimentary infill started under late-Glacial, low sea level conditions, with the deposition of fluviatile gravels and coarse sands. During marine transgression (8500–3000 cal years BP), these coarse sediments were trapped in the upper- and middle-estuarine areas by the landward migrating estuarine deposits, whereas in the lower estuary former fluviatile materials were reworked and included into this transgressive interval. Great volumes of mainly near-marine sediments were deposited in the lower estuary, alternating brackish and near-marine sediments were accumulated in the middle estuary, and brackish materials were sedimented in the upper estuary. The final part of the transgressive interval was represented by open-marine sediments limited above by the maximum flooding surface. The high-stand interval deposited during the upper Holocene (3000 cal years BP–nineteenth century human reclamation) represented brackish inter-tidal and supra-tidal conditions as the sedimentary infill was taking place under stabilized sea level.

However, since the mid-nineteenth century large-scale reclamation of inter-tidal areas has reduced the original estuary to a simple tidal channel. The discharge of significant quantities of untreated industrial and domestic effluents directly into the estuary led to the disappearance of the estuarine microfauna and to the significant pollution of the sediments. Three different zones could be identified in the recent sedimentary record, reflecting initially the pre-industrial estuarine conditions with abundant and diverse foraminiferal assemblages and baseline levels of metals, followed by the industrial period sedimentary record, when high concentrations of metals in the estuarine environment allowed the development of abundant foraminiferal assemblages in the older industrial zone (1850–1950) that disappeared in the newer industrial zone (1950–2000) due to complete defaunation of the Bilbao Estuary caused by minimum oxygen levels during this period. Obtained data provide important information in planning the definitive regeneration of the Bilbao Estuary, because sediments have proven to be an important storage reservoir for pollutants and microfossils.

Combined geochemical and microfaunal monitoring programs adequately reflect sediment quality in heavily polluted environments, making them suitable for assessing remediation schemes in estuarine systems. The significant reduction in contaminant sources and improvements in wastewater treatment have decreased pollutant disposal to the estuarine ecosystem, most notably since the WWTP implemented biological treatment in 2001. Metal concentrations and foraminiferal contents in surficial sediments reflect this improvement in the general environmental quality. However, there has been a significant delay in recovery as regards the expectations of the local water authority. In 2006, elevated pollutant concentrations remained widespread throughout the middle and lower estuary and the present day pollution status of this area is still cause for concern. Once the anthropogenic inputs dramatically decreased, pollutants accumulated in the sediments became a major problem. In fact, the environmental recovery of the Bilbao Estuary is strongly affected by the resuspension of the high concentration of pollutants stored in the sediments below the surface and, although less significantly, frequent spills from the WWTP.

References

Allen, George P., and Henry W. Posamentier. 1993. "Sequence Stratigraphy and Facies Model of an Incised Valley Fill: The Gironde Estuary, France." *Journal of Sedimentary Petrology* 63, no. 3: 378–91.

Alve, Elisabeth. 1991. "Benthic Foraminifera in Sediment Cores Reflecting Heavy Metal Pollution in Sørfjord, Western Norway." *Journal of Foraminiferal Research* 21, no. 1: 1–19.

———. 1999. "Colonization of New Habitats by Benthic Foraminifera: A Review." *Earth-Science Reviews* 46, nos. 1–4: 167–85.

Andersen, Jesper H., Daniel J. Conley, and Søren Hedal. 2004. "Palaeoecology, Reference Conditions and Classification of Ecological Status: The EU Water Framework Directive in Practice." *Marine Pollution Bulletin* 49, no. 4: 283–90.

Bald, Juan, Ángel Borja, Iñigo Muxika, Javier Franco, and Victoriano Valencia.

2005. "Assessing Reference Conditions and Physico-chemical Status According to the European Water Framework Directive: A Case Study from the Basque Country (Northern Spain)." *Marine Pollution Bulletin* 50, no. 12: 1508–22.

Bartolomé Luis, Itziar Tueros, E. Cortazar, Juan Carlos Raposo, Jon Sanz, Olatz Zuloaga, Alberto de Diego, Nestor Etxebarria, Luis ÁngelFernández, and Juan Manuel Madariaga. 2006. "Distribution of Trace Organic Contaminants and Total Mercury in Sediments from the Bilbao and Urdaibai Estuaries (Bay of Biscay)." *Marine Pollution Bulletin* 52, no. 9: 1090–1117.

Bero, Andrew S., and Ronald J. Gibbs. 1990. "Mechanisms of Pollution Transport in the Hudson Estuary." *Science of the Total Environment* 97/98: 9–22.

Borja, Ángel. 2005. "The European Water Framework Directive: A Challenge for Nearshore, Coastal and Continental Shelf Research." *Continental Shelf Research* 25, no. 14: 1768–83.

Borja, Ángel, Javier Franco, and V. Pérez. 2000. "A Marine Biotic Index to Establish the Ecological Quality of Soft-bottom Benthos within European Estuarine and Coastal Environments." *Marine Pollution Bulletin* 40, no. 12: 1100–14.

Borja, Ángel, Javier Franco, Victoriano Valencia, Adolfo Uriarte, and Raúl Castro. 2002. *Red de vigilancia y control de la calidad de las aguas litorales del País Vasco: otoño 2000–verano 2001.* UTE AZTI-LABEIN and Departamento de Ordenación del Territorio, Vivienda y Medio Ambiente del Gobierno Vasco. Unpublished report.

Borja, Ángel, Ibon Galparsoro, Oihana Solaun, Iñigo Muxika, Eva María Tello, Adolfo Uriarte, and Victoriano Valencia. 2006. "The European Water Framework Directive and the DPSIR: A Methodological Approach to Assess the Risk of Failing to Achieve Good Ecological Status." *Estuarine Coastal Shelf Science* 66, nos. 1–2: 84–96.

Casado-Martínez, M. Carmen, José Luis Buceta, Mª Jesús Belzunce, and T. Ángel DeValls. 2006. "Using Sediment Quality Guidelines for Dredged Material Management in Commercial Ports from Spain." *Environment International* 32, no. 3: 388–96.

Cearreta, Alejandro. 1988. "Distribution and Ecology of Benthic Foraminifera in the Santoña Estuary, Spain." *Revista Española de Paleontología* 3: 23–38.

———. 1998. "Holocene Sea-level Change in the Bilbao Estuary (North Spain): Foraminiferal Evidence." *Micropaleontology* 44, no. 3: 265–76.

Cearreta, Alejandro, Mª Jesús Irabien, Eduardo Leorri, Iñaki Yusta, Ian W. Croudace, and Andrew B. Cundy. 2000. "Recent Anthropogenic Impacts on the Bilbao Estuary, Northern Spain: Geochemical and Microfaunal Evidence." *Estuarine, Coastal and Shelf Science* 50, no. 4: 571–92.

Cearreta, Alejandro, Mª Jesús Irabien, Eduardo Leorri, Iñaki Yusta, A. Quintanilla, and Ane Zabaleta. 2002. "Environmental Transformation of the Bilbao Estuary, N. Spain: Microfaunal and Geochemical Proxies in the Recent Sedimentary Record." *Marine Pollution Bulletin* 44, no. 6: 487–503.

Consorcio de Aguas del Gran Bilbao. 1993. *Memoria 1992*. Bilbao: Consorcio de Aguas.

Cotano, Unai, and Fernando Villate. 2006. "Anthropogenic Influence on the Organic Fraction of Sediments in Two Contrasting Estuaries: A Biochemical Approach." *Marine Pollution Bulletin* 52, no. 4: 404–14.

Croudace, Ian W., and Andrew B. Cundy. 1995. "Heavy Metal and Hydrocarbon Pollution in Recent Sediments from Southampton Water, Southern England: A Geochemical and Isotopic Study." *Environmental Science and Technology* 29, no. 5: 1288–96.

Cundy, Andrew B., Ian W. Croudace, John Thomson, and James T. Lewis. 1997. "Reliability of Salt Marshes as 'Geochemical Recorders' of Pollution Input: A Case Study from Contrasting Estuaries in Southern England." *Environmental Science and Technology* 31, no. 4: 1093–1101.

Díez, Isabel, Alberto Santolaria, and José María Gorostiaga. 2003. "The Relationship of Environmental Factors to the Structure of Subtidal Seaweed Vegetation of the Western Basque coast (N Spain)." *Estuarine Coastal Shelf Science* 56, nos. 5–6: 1041–54.

Directive 200/60/EC of the European Parliament and of the Council of 23 October 2000 establishing a framework for Community action in the field of water policy. OJEC 2000; L 327. At http://eur-lex.europa.eu/LexUriServ/LexUriServ.do?uri=OJ:L:2000:327:0001:0072:EN:PDF.

García-Barcina, José María, José Antonio González-Oreja, and Alejandro de la Sota. 2006. "Assessing the Improvement of the Bilbao Estuary Water Quality in Response to Pollution Abatement Measures." *Water Research* 40, no. 5: 951–60.

García-Barcina, José María, Mikel Oteiza, and Alejandro de la Sota. 2002. "Modelling the Faecal Coliform Concentration in the Bilbao Estuary." *Hydrobiologia* 475–476: 213–19.

García-Merino, Luis Vicente. 1987. *La formación de una ciudad industrial. El despegue urbano de Bilbao*. Bilbao: Instituto Vasco de la Administración Pública.

González-Oreja, José Antonio, and José Ignacio Sáiz-Salinas. 1999. "Loss of Heterotrophic Biomass Structure in an Extreme Estuarine Environment." *Estuarine Coastal Shelf Science* 48, no. 3: 391–99.

Gorostiaga, José María, and Isabel Díez. 1996. "Changes in Sublittoral Benthic Marine Macroalgae in the Polluted Area of Abra de Bilbao and Proximal Coast (Northern Spain)." *Marine Ecology Progress Series* 130: 157–67.

Hazera, Jean. 1968. *La región de Bilbao et son arrière pays: Etude géomorphologique*. San Sebastián: Sociedad de Ciencias Naturales Aranzadi.

Irabien, Ma. Jesús. 1993. "Mineralogía y geoquímica de los sedimentos actuales de los ríos Nervión-Ibaizabal, Oka, Butrón y Nive. Índices de gestión ambiental." PhD diss., University of the Basque Country (UPV/EHU).

Irabien, Ma. Jesús, Iñaki Yusta, and Ane Zabaleta. 2001. "Metales pesados en sedimentos contaminados: variabilidad espaciotemporal en la ría de Bilbao." *Geogaceta* 30, 203–6.

Junta de Obras del Puerto de Bilbao. 1910. *Churruca y el Puerto de Bilbao (Apuntes biográficos, Antecedentes históricos y Resumen descriptivo de las obras de mejora de la Ría y Puerto de Bilbao).* Bilbao: Imprenta Emeterio Verdes.

Landajo, Amaia, Gorka Arana, Alberto de Diego, Nestor Etxebarria, Olatz Zuloaga, and David Amoroux. 2004. "Analysis of Heavy Metals Distribution in Surficial Estuarine Sediments (Estuary of Bilbao, Basque Country) by Open-focused Microwave-assisted Extraction and ICP-OES." *Chemosphere* 56, no. 11: 1033–41.

Leorri, Eduardo, and Alejandro Cearreta. 2004. "Holocene Environmental Development of the Bilbao Estuary, Northern Spain: Sequence Stratigraphy and Foraminiferal Interpretation." *Marine Micropaleontology* 51, nos. 1–2: 75–94.

Leorri, Eduardo, Alejandro Cearreta, Mª Jesús Irabien, and Iñaki Yusta. 2008. "Geochemical and Microfaunal Proxies to Assess Environmental Quality Conditions during the Recovery Process of a Heavily Polluted Estuary: The Bilbao Estuary Case (N. Spain)." *Science of the Total Environment* 396, no. 1: 12–27.

Liang, Yan, and Ming Hung Wong. 2003. "Spatial and Temporal Organic and Heavy Metal Pollution at Mai Po Marshes Nature Reserve, Hong Kong." *Chemosphere* 52, no. 9: 1647–58.

Long, Edward R., Donald D. MacDonald, Sherri L. Smith, and Fred D. Calder. 1995. "Incidence of Adverse Biological Effects within Ranges of Chemical Concentrations in Marine and Estuarine Sediments." *Environmental Management* 19, no. 1: 81–97.

McCauley, Dennis J., G. M. DeGraeve, and Tyler K. Linton. 2000. "Sediment Quality Guidelines and Assessment: Overview and Research Needs." *Environmental Science Policies* 3, suppl. 1: 133–44.

Muxika, Iñigo, Ángel Borja, and Juan Bald. 2007. "Using Historical Data, Expert Judgement and Multivariate Analysis in Assessing Reference Conditions and Benthic Ecological Status, According to the European Water Framework Directive." *Marine Pollution Bulletin* 55, nos. 1–6: 16–29.

Nayar, Sasi, Beverly P.L. Goh, Loke Ming Chou, and S. Reddy. 2003. "In Situ Microcosms to Study the Impact of Heavy Metals Resuspended by Dredging on Periphyton in a Tropical Estuary." *Aquatic Toxicology* 64, no. 3: 293–306.

Pirazzoli, Paolo A. 1991. *World Atlas of Holocene Sea-level Changes.* Amsterdam: Elsevier.

Puertos del Estado. 1997. *Atlas Geotécnico del Puerto de Bilbao.* Madrid: Centro de Publicaciones del Ministerio de Fomento.

Quintino, Victor, Michael Elliott, and Ana Maria Rodrigues. 2006. "The Derivation, Performance and Role of Univariate and Multivariate Indicators of

Benthic Change: Case Studies at Differing Spatial Scales." *Journal of Experimental Marine Biology and Ecology* 330, no. 1: 368–82.

Rivas, Victoria, and Antonio Cendrero. 1991. "Use of Natural and Artificial Accretion on the North Coast of Spain: Historical Trends and Assessment of Some Environmental and Economic Consequences." *Journal of Coastal Research* 7, no. 2: 491–507.

Ruiz, José Miguel, and José Ignacio Sáiz-Salinas. 2000. "Extreme Variation in the Concentration of Trace Metals in Sediments and Bivalves from the Bilbao Estuary (Spain) Caused by the 1989–90 Drought." *Marine Environment Research* 49: 307–17.

Sáiz-Salinas, José Ignacio. 1997. "Evaluation of Adverse Biological Effects Induced by Pollution in the Bilbao Estuary (Spain)." *Environmental Pollution* 96, no. 3: 351–59.

Sáiz-Salinas, José Ignacio, Gabriela Francés-Zubillaga, and Xabier Imaz-Eizaguirre. 1996. *Uso de bioindicadores en la evaluación de la contaminación de la Ría de Bilbao*. Bilbao: Servicio Editorial de la Universidad del País Vasco.

Sáiz-Salinas, José Ignacio, and José Antonio González-Oreja. 1998. "Coprostanol Levels and Organic Enrichment in Sediments of the Bilbao Estuary (North of Spain)." *Oceanologica Acta* 21, no. 2: 319–24.

Salas, Fuensanta, Concepción Marcos, João Miguel Neto, Joana Patrício, Ángel Pérez-Ruzafa, and João Carlos Marques. 2006. "User-friendly Guide for Using Benthic Ecological Indicators in Coastal and Marine Quality Assessment." *Ocean and Coastal Management* 49, nos. 5–6: 308–31.

Sharifi, Abdol Rahim, Ian W. Croudace, and R. L. Austin. 1991. "Benthic Foraminiferids as Pollution Indicators in Southampton Water, Southern England, U.K." *Journal of Micropalaeontology* 10, no. 1: 109–13.

Stronkhorst, Joost, Jos Brils, John Batty, Marina Coquery, Mike Gardner, Jaakko Mannio, Ciaran O'Donnell, Jaap Steenwijk, and Paul Frintorp. 2004. *Discussion Document on Sediment Monitoring Guidance for the EU Water Framework Directive*. Version 2. EU Water Framework Directive Expert Group on Analysis and Monitoring of Priority Substances. At www.sednet.org/download/AMPS_sediment_monitoring_discussion_doc_v2.pdf.

Uriarte, Ibon, and Fernando Villate. 2004. "Effects of Pollution on Zooplankton Abundance and Distribution in Two Estuaries of the Basque Coast (Bay of Biscay)." *Marine Pollution Bulletin* 49, no. 3: 220–28.

Urrutia, Javier. 1986. "Estudio de la estructura y funcionamiento del estuario del Nervión en relación a la dinámica del fitoplancton." PhD diss., University of the Basque Country (UPV/EHU).

Valette-Silver, Nathalie J. 1993. "The Use of Sediment Cores to Reconstruct Historical Trends in Contamination of Estuarine and Coastal Sediments." *Estuaries* 16, no. 3: 577–88.

6

The Landscape of the Autonomous Community of the Basque Country: The Evolution of Forest Systems

LORENA PEÑA and IBONE AMEZAGA

The University of the Basque Country (UPV/EHU)

In recent decades, landscapes have been changing at an unprecedented rate. The main agent of the change is anthropogenic in origin (Winged et al. 2004; Seto and Fragkias 2005; Plieninger 2006), leading to drastic changes that are in many cases difficult to reverse (Rutledge and Lepczyk 2002). The speed of change depends mainly on the intensity and frequency of disturbances (Liu and Taylor 2002). Often, landscapes form a mosaic that reflects the alteration of natural forests, forest plantations, pastures, crops, shrubs, and urban areas (Rigueiro 1999). The spatial distribution of the tiles (the "puzzle pieces" that form this mosaic) in the landscape is critical for the regulation of flows of matter and energy, such as for the dispersal of plant and animal species, as well as the functioning of its ecosystems (Johnson et al. 1992; Forman 1995; McGarigal and Marks 1995; Rutledge and Lepczyk 2002; Onaindia et al. 2004; Rodriguez et al. 2005; Martinez, Mokady, and Wool 2005).

One of the most important environmental problems today is the loss of biodiversity, due mainly to changes in land use (Vitousek et al. 1997; Sala et al. 2000; Rodriguez et al. 2005) and competition brought by invasive species (Whigham 2004). Urbanization, deforestation, industrialization, and intensive agriculture have resulted in rapidly changing landscapes (Forman 1995; Gurrutxaga 2004; Crow 2005). Historically, damage occurring to valued cultural landscape has resulted in the losses of ecological capacities within the ecosystem, in diversity, and in scenic beauty. Changes in the use of land have resulted primarily in habitat loss and increased isolation within the forest

* The following organizations that funded this research: the Spanish Ministry of Science and Innovation (project/grant CGL2008-05579-C02-01/BOS), and the Basque Department of Education, Universities, and Research (funding of research groups convening in 2009).

mosaic (Henle et al. 2004). In much of Europe (and also in the Middle East, the Mediterranean, and northern Africa) most of this transformation has occurred only in the last century (Wilcove, McLellan, and Dobson 1986).

Fragmentation of ecosystems is one of the spatial processes that generate further changes in the structure of landscapes in temperate zones. Moreover, it is probably one of the greatest threats to biodiversity (Turner et al. 1996; Harrison and Bruna 1999; Telleria and Santos 2000; Altamirano, Echeverría and Lara 2007). The fragmentation of the ecosystem is commonly associated with a reduction of the surface of a tile (Larsson 2001; Lopez-Barrera 2004); increased isolation (Forman 1995; Onaindia et al. 2000); and an accentuation of borders, or the *edge effect* (Andrén 1994; Telleria and Santos 2000; Larsson 2001; Gurrutxaga 2003, Fahrig 2003), which may adversely affect the biota that live there (Saunders, Hobbs, and Margules 1991; Murcia 1995; Yanez et al. 1999; Burel and Baudry 2003). The edge effect can be defined as the result of the interaction of two adjacent ecosystems (Murcia 1995; Lopez-Barrera 2004). The susceptibility of forest fragments to harmful effects at its borders is one of the most obvious consequences of the fragmentation of forest ecosystems (Forman 1995; Lopez-Barrera 2004; Rodriguez 2007). This is because of the higher degree of contrast between two habitats, specifically the lower the biological flows (movement of organisms between habitats) and the increased physical flows (lateral light penetration into the forest) (Weathers, Cadenasso, and Pickett 2001; Lopez-Barrera 2004).

Another important consequence of fragmentation is the isolation of forest tile, which interferes with the movement of individual species between adjacent ecosystems of differing quality. Before fragmentation occurs, species maintain certain populations due to the contribution of individuals within that species and seeds from the most dense and highest quality populations. Nevertheless, after the fragmentation occurs, these inputs end and the difficulties in getting these seeds increase and may cause a cascade of local extinctions in suboptimal fragments. In reality, high connectivity can favor local populations through immigration, as explained in the theory of *metapopulations*.

Although today it is believed that Mediterranean and tropical areas are more vulnerable to the effects of mild fragmentation (Henle et al. 2004), its negative consequences are evident in forest ecosystems within temperate zones that have been altered in both contemporary and historical contexts (Whitcomb et al. 1981; Harris 1984; Wilcove, McLellan, and Dobson 1986; Santos, Telleria and Virgos 1999; Santos, Telleria, and Carbonell 2002; Lindenmayer and Franklin 2002; Garcia et al. 2005). A clear example of this is the changing landscape in the CAPV-EAE. For centuries, the landscape of the CAPV-EAE underwent a traditional agroforestry–livestock model of cultivation in which in the village hamlet played a very important role. That is, its main purpose was self-sufficient consumption, which supported a wide variety of crops and harvests that made for a high-diversity landscape (Michel 2003; Rodriguez 2007). However, in the late-nineteenth and early-twentieth centuries the growth of industry in the Cantabrian slope of the CAPV-EAE (i.e., in the provinces of Bizkaia and Gipuzkoa) led to the relocation of most rural people to urban centers. Consequently, the progressive abandonment of iron works significantly reduced the demand for charcoal. It was at this juncture that the CAPV-EAE reached its historical minimum of forested area, due to the by then uncontrolled consumption of forest ecosystems for their supply of charcoal. In fact, the situation in Bizkaia and Gipuzkoa was so untenable that both provincial councils established a forest service, whose first mission was repopulation (Groome 1990).

The introduction of *Pinus radiata* (the Monterey pine from California in the United States) in the late-nineteenth century by Mario Adan Ayarza created great expectations regarding the recovery of forests for the future through the establishment of native vegetation (e.g., mixed forest *Quercur robur*, *Castanea sativa*, etc.) (Michel 2003). However, high profit was gleaned from this exotic species and a subsidy policy ensued. Moreover, in Bizkaia and Gipuzkoa, most forestland belonged to private owners who favored the establishment of this species. This repopulation, which replaced both crop and forests with native hardwood species, began to cover forest areas, forming artificial landscapes that were a result of the fact that these plantations were geared exclusively toward production and private profit. This repopulation effort diminished, however, with the frost of 1956, which destroyed much of the young plantations of *Pinus radiata* in Gipuzkoa (Michel 2003, 157–58). This led to the diversification of other introduced species, such as larch, Corsican pine, and Douglas fir.

The Mediterranean side of the CAPV-EAE suffered less deforestation. On the one hand, this was because there was a lower population density and, therefore, less industrial development. On the other, a higher proportion of common land or public forestland existed that was better maintained. In 1984 in the CAPV-EAE, 93.7 percent of production was pine and only 2 percent was oak, beech, and chestnut. Ten years later, the production of oak, beech, and chestnut rose to 8.51 percent (versus 91.48 percent of pine). Between 1984 and 1993, the timber market suffered a major crisis: the importation of very cheap wood from other countries such as the United States, Chile, and Russia. This crisis was coupled with a paper-mill crisis, with mills being the main consumers of wood produced in the CAPV-EAE. The paper-mill crisis led to the closure of many mills as the demand for pine dropped considerably (Michel 2003, 215–16). At present, conifers provide 63 percent of the total volume of wood, compared to 67 percent nine years before.

In the 1990s, the CAPV-EAE became more aware of environmental problems. This is evident in the Basque Government's *Forest Plan for 1994 to 2030*, which states:

> The protection of soil and water resources, the conservation of biodiversity (and of those sites where human activities possess an undeniable cultural value), the natural architecture of the landscape, the enhancement of unused resources, and the necessary supply of material goods must be integrated into an equation that meets the physical and human conditions of the territory (Nd, 119).

Given the importance of analyzing the evolution of the landscape in land management and in the assessment of the *Forest Plan*, this study seeks to evaluate the effects of forest change in the current landscape structure, especially in potential vegetation dynamics (i.e., natural forests) between 1996 and 2005. This analysis is practicable for assessing the impact of environmental legislation and management to facilitate, as required, new strategies both in law and management.

Methodology

The databases used for analysis were obtained from digitized forest maps of the CAPV-EAE in 1996 and 2005 and made available by the Basque Government's IKT (*Nekazal Ikerketa eta Teknologia* [Agricultural Research and Technology Institute]). These databases have been managed using geographic information systems (GIS). The landscape of the CAPV-EAE was classified into three major land uses, within which other uses include nonforested areas consisting of grasslands, crops, and shrubs, as well as plantations comprised of conifers, eucalyptus, and hardwoods. Also included are natural forests comprised of oak (including cork oak, *Quercus suber*), gall, mixed-forest oak, beech, and birch, as well as Pyrenean oak (including *Quercus pyrenaica* and *Quercus pubescens*), riparian, and cliff forests. The discussion that follows begins with the databases developed through the vLATE (vector-based landscape analysis tools extension) program (Lang and Tiede 2003).[1]

Average Distance to the Nearest Tile Index

An ADNT index (average distance to nearest tile) shows the average distance in meters from the edge of a tile to the edge of the nearest tile that has the

1. These data were then inserted into ArcGIS 9.0, and using the formulation proposed in the program FRAGSTATS (McGarigal et al. 2002) to calculate landscape rates of change (Santos, Telleria, and Virgos 2002; Garcia et al. 2005; Watson, Whittaker, and Freudenberger 2005).

same land usage, which gives an idea of the degree of isolation that the tile is exposed to.

Fragmentation Index

A fragmentation index (F) is a mapping measure that shows the degree of spatial aggregation of the tile habitat under study. It uses a level inversely proportional to the degree of fragmentation. An increase in the index value is related to the degree of fragmentation and vice versa. Therefore, increasing the fragmentation means a decrease in the total area, as well as an increase in the number and dispersion of the same tile (Gurrutxaga 2003).

Fragmentation indices were calculated at the level for each tile, at the level of vegetation, and at the landscape level. In order to get a complete picture of the change in the landscape, it is important to examine the rates of change of the landscape at different levels. Equally important, these indices have calculated the area and the number of tiles, in addition to the percentage of the landscape's area with respect to total area and average size (Terradas et al. 2004; Watson, Whittaker, and Dawson 2004; Rodriguez 2007). Kurtis Trzcinski, Lenore Fahrig, and Merriam Gray (1999) show that independent measures of fragmentation can be derived using both the number of tiles in the landscape, as well as their size.

Statistical Treatment

In order to study the evolution of the landscape between 1996 and 2005 (in addition to the calculation of landscape indices), both the ADNT and F indices were compared between the two years of study for the different types of tile that form the landscape of the CAPV-EAE, as well as for the different types of natural forests via the nonparametric Mann-Whitney U test (after verifying that the data were not normal and could not be normalized).

Results

The Landscape Evolution of the CAPV-EAE between the Years 1996 and 2005

Between 1996 and 2005, the landscape structure of the CAPV-EAE saw a change in land use on 139,440 ha, which is equivalent to 19.3 percent of its total surface (table 6.1). New nonforested areas (derived from the processing of 3,809 ha of mixed oak forest and 20,003 ha of conifers) comprise 23.7 percent (33,056 ha) of the surface change. Thirty percent (41,631 ha) is new forest plantations (i.e., pine, eucalyptus, and hardwoods mainly from nonforested areas). The rest are new areas occupied by natural forests (46.3 percent), of which 26,540 ha is new surface occupied by mixed oak forest (which comes mainly from old plantations of conifers and old shrubs), 6,802 ha of gall, and

5,765 ha of oak. In addition, an important fact to note is that the urban area has increased currently by 32 percent since 1996. Thus, the CAPV-EAEs present surface area is approximately 42,000 ha.

Table 6.1. Value of total area and its percentage in respect to the total area and the forest area occupied for each of the land uses that form part of the 2-year study within the CAPV-EAE.

	Area (ha)		% Total area		% Forest area	
	1996	2005	1996	2005	1996	2005
Nonforested areas	361,677	338,696	50.04	46.87		
Conifer plantations	203,068	193,367	28.10	26.76	56.24	50.37
Eucalyptus plantations	9,088	11,498	1.26	1.59	2.52	2.99
Hardwood plantations	3,709	4,440	0.51	0.61	1.03	1.16
Riparian forests	3,167	4,790	0.44	0.66	0.88	1.25
Oak	24,694	28,845	3.42	3.99	6.84	7.51
Gall	20,413	24,604	2.82	3.40	5.65	6.41
Mixed oak forests	34,248	48,896	4.74	6.77	9.48	12.74
Beech	50,647	54,759	7.01	7.58	14.03	14.26
Birch	114	122	0.02	0.02	0.03	0.03
Pyrenean oak	11,539	12,331	1.60	1.71	3.20	3.21
Cliff forests	418	247	0.06	0.03	0.12	0.06

The landscape, as seen in the number of tiles present in the area, has remained more or less constant, declining only by 0.4 percent (from 10,060 tiles in 1996 to 10,017 tiles in 2005). However, the average sizes of the tiles were significantly lower in 2005. This did not include an increase in isolation of tiles (table 6.2). As for the isolation of the tiles, it is generally observed that the average distance to the nearest tile of same usage has not changed significantly. Together with the fragmentation index, the ADNT measure has remained fairly constant in recent years and suggests that the fragmentation of the CAPV-EAE landscape, in general, has not changed over the period observed. However, it is necessary to analyze whether this fragmentation has been maintained between all uses of the landscape or has changed due to certain applications.

Evolution of Different Land Uses

In the case of natural forests, there is an increase of 20.2 percent in the area covered by forest ecosystems (from occupying a surface of 145,240 ha in 1996 to occupying an area of 174,595 ha in 2005). This contrasts with what occurs in both nonforested areas and forest plantations, which suffer a decline in their surfaces of 6.35 percent and 3.04 percent, respectively (table 6.3 and figure 6.1a). Despite these surface decreases, the number of tiles of each type of usage

Table 6.2. Calculated index values (Average ± SE) for all of the tiles that make up the landscape of the BAC (within the two-year study), and the degree of significance for the nonparametric Mann-Whitney U. The abbreviations correspond to AT: average size of the tiles; ADNT: average distance to the nearest tile of same usage; F: fragmentation index; p: level of significance.

Year	AT (ha)	ADNT (m)	F
1996	71.85 ± 25.86	344.26 ± 10.72	271.17
2005	52.34 ± 5.820	349.84 ± 9.850	274.94
p	0.000**	0.250	

* $p \leq 0.05$
** $p \leq 0.01$
*** $p \leq 0.001$

has not increased or decreased proportionately. Thus, only natural forests have increased by 8 tiles, from 4,816 in 1996 to 4,824 tiles in 2005. Nonforested areas have decreased by 119 tiles, which is equivalent to a decrease of 1.1 percent of the total area. Forest plantations have increased by 68 tiles, which represents an increase of 1 percent the total area (table 6.3 and figure 6.1b).

Table 6.3. A representation of the total area occupied by each of the types of land uses that have been classified within the tiles that make up the landscape of the CAPV-EAE (and the number of tiles that correspond to each within the two-year study).

	Area (ha)		Number of tiles	
	1996	2005	1996	2005
Nonforested areas	361,677	338,696	2.085	1.966
Forest plantations	215,861	209,305	3.159	3.227
Natural forests	145,240	174,595	4.816	4.824

Figure 6.1. A representation of the percentages of the area occupied by each type of use where the classified tiles that shape the landscape of the CAPV-EAE (and the percentage of same tiles present within the two years of study).

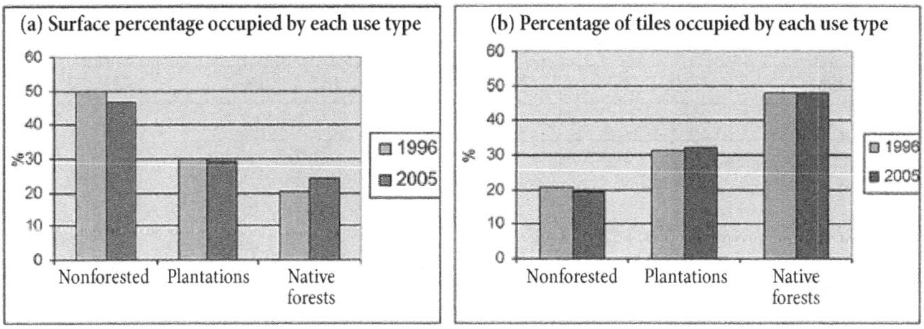

The landscape-level analysis shows that differences (i.e., diverse types of uses) within the classified tiles are significant in all cases, except for the average distance to the nearest tile of the same usage. This is significantly higher only among forest plantations ($p \leq 0.001$) (table 6.4). Thus, forest plantations are more isolated in 2005 than in 1996, as its tiles are more distant from each other.

As for the average size of the tiles (including natural forests tiles), there was a significant increase ($p \leq 0.001$), from an average size of 30 ha in 1996, to 36 ha in 2005. This contrasts to what happens with nonforested areas and forest plantations, which have suffered a significant decline of 0.7 percent and 5.1 percent in their average size, respectively ($p < 0.05$) (table 6.4). However, with respect to the degree of fragmentation, there was an increase among forest plantations and a slight decline among nonforested areas and natural forests, which are consistent with the values obtained for the nearest distance to the tile of the same usage (table 6.4).

Table 6.4. Calculated index values (Average ± SE) for the different land uses that have been classified within the tiles that make up the landscape of the CAPV-EAE within the 2-year study (and the degree of significance for the nonparametric Mann-Whitney U). The abbreviations correspond to AT: average size of the tile; ADNT: average distance to the nearest tile of same usage; F: fragmentation index; p: level of significance.

	Nonforested areas			
Year	AT (ha)	A-SHAPE	ADNT (m)	F
1996	173.5 ± 123.2	1.883 ± 0.034	131.58 ± 3.12	717.88
2005	172.3 ± 101.5	2.138 ± 0.039	136.68 ± 3.76	727.73
p	0.002**	0.000**	0.201	
	Plantations			
Year	AT (ha)	A-SHAPE	ADNT (m)	F
1996	68.33 ± 12.41	1.82 ± 0.02	289.53 ± 11.94	132.83
2005	64.86 ± 12.74	1.90 ± 0.02	342.53 ± 12.45	116.94
p	0.024*	0.000**	0.000**	
	Natural forests			
Year	AT (ha)	A-SHAPE	ADNT (m)	F
1996	30.16 ± 2.61	2.035 ± 0.012	472.2 ± 20.7	15.06
2005	36.19 ± 3.46	2.383 ± 0.017	441.4 ± 18.47	19.29
p	0.000**	0.000**	0.122	

* $p \leq 0.05$

Evolution of Natural Forests Tiles

All natural forests (except cliff forests, whose initial surface area decreased by 41 percent), have had surface increases since 1996. The largest increases are among the riparian forest, with a 51.3 percent surface increase in 1996. This is followed by the mixed oak forest (42.8 percent increase), and then gall (20.5 percent increase) (table 6.1). Regarding the number of tiles, it should be noted that only riparian forests and oak forests have seen an increase in number since 1996, whereas the rest of the natural forests saw a reduction (table 6.5).

Table 6.5. Number of tiles occupied by different types of natural forests in the CAPV-EAE within the two-year of study. and the values of the percentages they occupy with respect to the total of the tiles and the tiles that make up the landscape of the CAPV-EAE. The represented values for beech are in blue.

	Number of tiles (NT)		% NT totals		% NT forested	
	1996	2005	1996	2005	1996	2005
Riparian forests	351	389	3.49	3.88	4.40	4.83
Oak	522	519	5.19	5.18	6.55	6.45
Gall	663	628	6.59	6.27	8.31	7.80
Mixed oak forests	2.238	2.345	22.25	23.41	28.06	29.13
Beech	781	734	7.76	7.33	9.79	9.12
Birch	16	14	0.16	0.14	0.20	0.17
Pyrenean oak	214	177	2.13	1.77	2.68	2.20
Cliff forests	31	18	0.31	0.18	0.39	0.22

At the landscape level, all types of natural forests have increased their average size from 1996 to 2005 by greater than 20 percent, except the oaks, which barely reach 18 percent, and cliff forests, which had the lowest recorded increase of 1.2 percent. However, these differences are significant only for riparian forests, mixed oak forests, and beech. In the case of isolation, riparian and oak forests are also unique because they reduced their distance to the nearest tile of the same usage. In the cliff forests, this distance increased significantly in 2005, indicating that these forests are further away in distance.

Table 6.6. Calcuated index values at the landscape level (Average ± SE) for the different types of natural forests that are part of the CAPV-EAE within the two-year study (and their degree of significance for the nonparametric Mann-Whitney U). The abbreviations correspond to AT: average size of the tiles; ADNT: average distance to the nearest tile of same usage; p: level of significance.

		1996	2005	
	Indices	Average ± SE	Average ± SE	p
Riparian forests	AT (ha)	9.020 ± 0.560	12.31 ± 0.710	0.000**
	ADNT (m)	1.199 ± 77.13	877.2 ± 52.81	0.015*
Oak	AT (ha)	47.31 ± 11.84	55.58 ± 12.84	0.057
	ADNT (m)	569.3 ± 51.10	529.2 ± 44.53	0.871

	Indices	1996 Average ± SE	2005 Average ± SE	p
Gall	AT (ha)	30.79 ± 2.820	39.18 ± 4.430	0.275
	ADNT (m)	276.6 ± 25.06	287.8 ± 17.79	0.684
Mixed oak forests	AT (ha)	15.30 ± 0.610	20.85 ± 0.800	0.000**
	ADNT (m)	306.9 ± 15.54	278.6 ± 10.95	0.001**
Beech	AT (ha)	64.85 ± 12.65	74.60 ± 19.76	0.019*
	ADNT (m)	347.7 ± 26.95	419.6 ± 34.63	0.073
Birch	AT (ha)	7.140 ± 0.860	8.730 ± 1.420	0.473
	ADNT (m)	7.273 ± 3000	9.837 ± 3.381	0.400
Pyrenean oak	AT (ha)	53.92 ± 18.07	69.67 ± 18.89	0.417
	ADNT (m)	938.9 ± 224.5	551.0 ± 71.09	0.787
Cliff forests	AT (ha)	13.49 ± 2.050	13.72 ± 3.880	0.619
	ADNT (m)	3.131 ± 955.8	7.571 ± 1.813	0.015*

* $p \leq 0.05$
** $p \leq 0.01$

In general, fragmentation decreases in all types of natural forests, except for cliff forests (table 6.7). This decrease is by more than 25 percent in all forests studied, except in beech and birch forests. In the territory of the CAPV-EAE, Pyrenean forests were the least fragmented, while riparian and oak forests were the most.

Table 6.7. Fragmentation index (F) calculated for the different types of natural forests that are part of the CAPV_EAE (within the two-year study).

	F	
	1996	2005
Riparian forests	24.34	40.96
Oak	180.7	229.7
Gall	190.6	246.1
Mixed oak forests	25.30	36.23
Beech	271.2	274.9
Birch	69.62	71.94
Pyrenean oak	304.7	810.9
Cliff forests	157.8	114.3

Discussion

In general, landscapes are constantly changing, often due to the growing influence of human beings. This is true in the case of the CAPV-EAE, where it was observed that 19.3 percent of the total land area has changed its land use. Although the territory of forest area has remained roughly constant, many of the tiles have changed in usage. For example, much of the area occupied by conifer plantations in 1996 has been cleared for logging. After being harvested, a lot of land has been abandoned, becoming thicket. Thus, the area once occupied by conifer plantations in 1996 was classified as nonforested areas in 2005, and, vice versa, conifer plantations now occupy much of the area classified in 1996 as nonforested areas. In the case of new tiles of mixed oak forest, they are derived primarily from old thicket and old conifer plantations. Thus, the substitution of nonforested areas of natural forest indicates a secondary succession, a pinnacle state compatible with its potential vegetation. In the CAPV-EAE, the typical potential vegetation in zones of higher elevation is populated by beech, whereas a descent in altitude leads to strips of mixed oak–dominated forests.

There has been an increase in artificial surface area in recent years due to urban development. Specifically, a large number of houses have been built and industry now occupies a significant part of the land. In fact, in 2000 residential homes accounted for 39 percent of the total floor area, economic activities 18 percent, and transport infrastructure 43 percent (Basque Government 2004). As urban areas expand, transform, and envelop the surrounding landscape, they adversely affect the natural environment at multiple spatial and temporal scales. This impact on the environment is changing weather conditions and the microclimate, causing habitat loss and fragmentation of biodiversity, due to the increased demand for natural resources (Seto and Fragkias 2005).

Evolution of Different Land Uses

Forest plantations showed less of an increase in their surface compared to natural forests. There has been a diversification of plantations. For example, eucalyptus and hardwood plantations now occupy part of the area occupied in 1996 by conifer plantations. This is the result of intensive forestry throughout the country, especially in Bizkaia and Gipuzkoa. Although this essay has not differentiated between the various species of conifers, according to the latest Forest Inventory of the Basque Government in 2006, there is a downward trend of the species *Pinus radiata* and an increase in other conifer species, such as the Douglas fir (*Pseudotsuga menziesii*).

In the case of hardwood plantations, repopulation generally occurs in publicly owned forests (Basque Government 2007). The species that is most introduced is the American oak (*Quercus rubra*), which has a short turnaround time of about sixty years (Association of Owners Gipuzkoa Forest 2003). Compared to the fast-growing species, the hardwood plantations provide environmental benefits such as contributing to soil stabilization and the development of wildlife communities, as well as increased biodiversity and landscape improvements. Yet, their proliferation has been limited so far, due to the lack of public policy support, relevant services, and subsidies. Since these kinds of plantations are concentrated mainly on public land, their development brings an economic burden to private landowners who do not have governmental support or subsidies.

Further, the numbers of *Pinus radiata* were reduced in favor of eucalyptus plantations. *Pinus radiata* mainly occupies land in Bizkaia, accounting for 56 percent of the forested area of this region, which is usually privately owned and represents 81 percent of the area occupied by this species (Basque Government 2006). Landowners use this pine because it has a shorter turnaround time, and thus a higher profitability (Forest Owners Association of Gipuzkoa 2003). Yet, in Bizkaia, after forest fires burned over 17,000 ha in 1989 (Galley and Saenz 1990), the species then most used in the reforestation was eucalyptus. This tree has a high growth rate and a high capacity to regrow after cutting or fire. In Gipuzkoa, however, the Douglas fir was most utilized (Michel 2006, 222).

Nonetheless, the upward trend of eucalyptus forests is troubling because it is an exotic species with a very short turnaround—even shorter than the thirty- to thirty-five-year turnaround of *Pinus radiata* (Forest Owners Association of Gipuzkoa 2003). Consequently, there is a strong imbalance between the nutrient extraction involved in logging (timber harvested and erosion losses) and the nutrient inputs; that is, the soil quality is becoming poorer. In addition, its high growth rate and transpiration significantly influence surface runoff in large areas, which decreases the regional water supply.

Therefore, reforestation planning should incorporate native species or otherwise use species with longer turnaround times (the estimated optional rotation age is tree trunks over 45 years old). Furthermore, it should favor the use of polycultures versus monocultures, as many experts advise (Sedjo and Botkin 1997; Donald et al. 1998; Stanley and Montagnini 1999; Hartley 2002). These arguments represent sustainable measures as reflected in the Basque Country's rural development program of 2007 to 2013. Such measures would conserve and regenerate forests with indigenous trees. They also seek to limit deforestation (caused by species with short-rotation shifts under thirty-five years) and potentiate the growth of species with medium- to low-rotation shifts.

In general, natural forests were the biggest benefactors during the period of 1996 to 2005, showing a broad tendency to increase in total area and average tile size. Thus, at a landscape level their fragmentation has decreased. This is gradually reflected in better quality vegetation patches, which promote their conservation and maintenance. The decrease of forest plantations at a landscape level is actually a positive factor, given the issues with exotic trees having short-rotation shifts. In general, the trees with longer shifts are more beneficial to the CAPV-EAE. In turn, the soil and water would have an ecological benefit as well as an equally positive impact on ecological services.

Evolution of Natural Forests Tiles

In general, every type of natural forest area studied increased as compared to 1996, except for cliff forests. These forests are usually found in various climates where soil conditions and climate are not optimal for the species of tree that make up this type of forest ecosystems (Aseginolaza et al. 1988). The oak forests in this area are composed of one variant of mixed hardwoods and are limited and constrained in development by the scarcity of land and poor soil. The only variant of beech found in this woodland is *Petrano calcicolous*.

Furthermore, oak forests have had a small recuperation. They have increased in surface and average tile size, while also reducing in isolation and degree of fragmentation. Oak and beech forests have the highest potential impact on vegetation in the CAPV-EAE. Oak, which is found at lower altitudes than the beech, is used less often in agriculture (Garcia et al. 2005) and forestry in the north mainly due to its accessibility (GESPLAN 2002). However, at present, it constitutes a breed of small-sized trees that are expanding (after the abandonment of agricultural land and the felling of ancient pines), despite suboptimal conditions.

Similar to oak forests, riparian forests have had a small recuperation. They have also increased in surface and average tile size, while reducing in isolation and degree of fragmentation. Yet, both of these forest ecosystems, despite their ecological importance and despite legislation, have been heavily impacted by the growth of far-ranging meadows, forest plantations, and crop planting that yields higher profits.

Conclusion

The landscape of the CAPV-EAE, in general, has experienced detectable changes due to land use. The changes have been manifest in decreases in the average size of tiles and increases in fragmentation. However, if examined more closely, these changes—contrary to the general trend—have actually decreased the fragmentation and increased the size of native vegetation tiles. These positive changes have been detected mainly in the tiles of oak and riparian forests, thereby encouraging their conservation and maintenance and enhancing their overarching importance to the landscape. Although oak remains the most fragmented vegetation, it could potentially dominate the landscape of the CAPV-EAE. Therefore, although current forestry policies seem to favor the recovery of native vegetation (given its benefits to society), it should not be forgotten that future conservation strategies for natural ecosystems should continue to be geared toward reducing the average distance between tiles of the same usage through the creation and recovery of ecological methodologies.

References

Alados, Concepción L., Yolanda Pueyo, Olivia Barrantes, Juan Escós, L. Giner, and Ana B. Robles. 2004. "Variations in Landscape Patterns and Vegetation Cover between 1957 and 1994 in a Semiarid Mediterranean Ecosystem." *Landscape Ecology* 19, no. 5: 543–59.

Altamirano, Adison, Cristian Echeverría, and Antonio Lara. 2007. "Efecto de la fragmentación forestal sobre la estructura vegetacional de las poblaciones amenazadas de *Legrandia concinna* (Myrtaceae) del centro-sur de Chile." *Revista chilena de historia natural* 80, no. 1: 27–42.

Andrén, Henrik. 1994. "Effects of Habitat Fragmentation on Birds and Mammals in Landscapes with Different Proportions of Suitable Habitat: A Review." *Oikos* 71: 355–66.

Aseginolaza, Carlos, Daniel Gómez, Xabier Lizaur, Gabriel Monserrat, Gregorio Morante, María R. Salaverria, and Pedro Ma. Uribe-Echebarria. 1988. *Vegetación de la Comunidad Autónoma del País Vasco.* Vitoria-Gasteiz: Servicio Central de Publicaciones, Gobierno Vasco.

Asociación de Propietarios Forestales de Guipúzcoa. 2003. *Basoberritzea eta basoaren erabilera: Repoblación y Manejo Forestal.* Donostia-San Sebastián: Departamento de Agricultura y Medio Ambiente de la Diputación Foral de Guipúzcoa.

Burel, Françoise, and Jacques Baudry. 2003. *Landscape Ecology: Concepts, Methods, and Applications.* Enfield, NH: Science Publishers.

Crow, Thomas R. 2005. "Landscape Ecology and Forest Management." In *Issues and Perspectives in Landscape Ecology*, edited by John A. Weins and Michael R. Moss. Cambridge: Cambridge University Press.

Donald, Paul F., Robert J. Fuller, A.D. Evans, and Susan J. Gough. 1998. "Effects of Forest Management and Grazing on Breeding Bird Communities in Plantations of Broadleaved and Coniferous Trees in Western England." *Biological Conservation* 85, nos. 1–2: 183–97.

Fahrig, Lenore. "Effects of Habitat Fragmentation on Biodiversity." 2003. *Annual Review of Ecology, Evolution and Systematics* 34: 487–515.

Forman, Richard T.T. 1995. "Some General Principles of Landscape and Regional Ecology." *Landscape Ecology* 10, no. 3: 133–42.

Galera, Antonio, and Daniel Saenz. 1990. "Los incendios forestales en Euskadi durante 1989, los más graves de su historia reciente." *Sustrai* 18: 46–50.

García, Daniel, Mario Quevedo, José R. Obeso, and Adán Abajo. 2005. "Fragmentation Patterns and Protection of Montane Forest in the Cantabrian Range (NW Spain)." *Forest Ecology and Management* 208, nos. 1–3: 29–43.

GESPLAN. *Sistema de cartografía de la C.A.P.V.* 2002. Vitoria-Gasteiz: Departamento de ordenación del territorio, vivienda y medio ambiente del Gobierno Vasco.

Gobierno Vasco. 2004. "Biodiversidad." In *Estado del medio ambiente en la CAPV*. Vitoria-Gasteiz: Gobierno Vasco.

———. 2006. *Inventario Forestal de la Comunidad Autónoma del País Vasco*. N.p: Gobierno Vasco.

———. N.d. *Plan Forestal Vasco 1994–2030*. N.p: Gobierno Vasco. At www.nasdap.ejgv.euskadi.net/r50-7393/es/contenidos/plan_programa_proyecto/plan_forestal/es_9394/adjuntos/Plan_forestal.pdf.

———. 2007. *Programa de Desarrollo Rural Sostenible del País Vasco 2007–2013: Repercusión de la anterior programación*. N.p.: Gobierno Vasco.

Groome, Helen. 1990. *Historia de la política forestal en el Estado español*. Monografías de la Agencia de Medio Ambiente de la Comunidad Autónoma de Madrid 1. Madrid: Agencia de Medio Ambiente.

Gurrutxaga, Mikel. 2003. *Índices de fragmentación y conectividad para el indicador de biodiversidad y paisaje de la Comunidad Autónoma del País Vasco*. Arkaute: IKT.

———. 2004. *Conectividad ecológica del territorio y conservación de la biodiversidad: Nuevas perspectivas en ecología del paisaje y ordenación territorio*. Vitoria-Gasteiz: Servicio Central de Publicaciones del Gobierno Vasco.

Harris, Larry D. 1984. *The Fragmented Forest: Island Biogeography Theory and the Preservation of Biotic Diversity*. Chicago: University of Chicago Press.

Harrison, Susan, and Emilio Bruna. 1999. "Habitat Fragmentation and Large-scale Conservation: What Do We Know for Sure?" *Ecography* 22, no. 3: 225–32.

Hartley, Mitschka J. 2002. "Rationale and Methods for Conserving Biodiversity in Plantation Forests." *Forest Ecology and Management* 155, nos. 1–3: 81–95.

Henle, Klaus, Kendi F. Davies, Michael Kleyer, Chris Margules, and Josef Settele. 2004. "Predictors of Species Sensitivity to Fragmentation." *Biodiversity and Conservation* 13, no. 1: 207–51.

Johnson, A.R., John A. Wiens, Bruce T. Milne, and Thomas O. Crist. 1992. "Animal Movement and Population Dynamics in Heterogeneous Landscapes." *Landscape Ecology* 7, no. 1: 63–75.

Lang, Stefan, and Dirk Tiede. 2003. *vLATE Extension für ArcGIS - vektorbasiertes Tool zur quantitativen Landschaftsstruktuanalyse*. CD-ROM. ESRI Anwenderkonferenz, Innsbruck.

Larsson, Tor-Björn (coordinator). 2001. *Biodiversity Evaluation Tools for European Forests*. Ecological Bulletins 50. Oxford: Blackwell Science.

Lindenmayer, David B., and Jerry F. Franklin. 2002. *Conserving Forest Biodiversity: A Comprehensive Multiscaled Approach*. Washington, D.C.: Island Press.

Liu, Jianguo, and William W. Taylor, eds. 2002. *Integrating Landscape Ecology into Natural Resource Management*. Cambridge: Cambridge University Press.

López-Barrera, Fabiola. 2004. "Estructura y función en bordes de bosques." *Ecosistemas* 13, no. 1: 67–77.

Martinez, Jean J., Ofer Mokady, and David Wool. 2005. "Patch Size and Patch Quality of Gall-inducing Aphids in a Mosaic Landscape in Israel." *Landscape Ecology* 20, no. 8: 1013–24.

McGarigal, Kevin, and Barbara J. Marks. 1995. *FRAGSTATS: Spatial Pattern Analysis Program for Quantifying Landscape Structure.* Portland, OR: USDA, Forest Service, Pacific Northwest Research Station.

McGarigal, Kevin, Sam Cushman, Maile Neel, and Eduard Ene. 2002. *FRAGSTATS: Spatial Pattern Analysis Program for Categorical Maps.* At www.umass.edu/landeco/research/fragstats/fragstats.html.

Michel, Mario. 2003. "El pino radiata (*Pinus radiata* D.Don) en la historia forestal de la Comunidad Autónoma del País Vasco: Análisis de un proceso de forestalismo intensivo." PhD diss., Polytechnic University of Madrid.

——. 2006. *El pino radiata en la historia forestal vasca: Análisis de un proceso de forestalismo intensivo.* Supplement of *Munibe.* Donostia-San Sebastián: Sociedad de Ciencias Aranzadi.

Murcia, Carolina. 1995. "Edge Effects in Fragmented Forests: Implications for Conservation." *Trends in Ecology and Evolution* 10, no. 2: 58–62.

Onaindia, Miren, Ibone Amezaga, Ander González, and Izaskun Domínguez. 2000. *Evaluación de la biodiversidad de los ecosistemas forestales de Bizkaia.* Unpublished.

Onaindia, Miren, Izaskun Domínguez, Isabel Albizu, Carlos Garbisu, and Ibone Amezaga. 2004. "Vegetation Diversity and Vertical Structure as Indicators of Forest Disturbance." *Forest Ecology and Management* 195, no. 3: 341–54.

Plieninger, Tobias. 2006. "Habitat Loss, Fragmentation, and Alteration—Quantifying the Impact of Land-use Changes on a Spanish Dehesa Landscape by Use of Aerial Photography and GIS." *Landscape Ecology* 21, no. 1: 91–105.

Rigueiro, Antonio. 1999. "El paisaje en la gestión sostenible de los bosques." In *Actas del Congreso de Ordenación y Gestión Sostenible de Montes (Santiago de Compostela, 4–9 de octubre de 1999).* Volume 2. Santiago de Compostela: Conselleria de Medio Ambiente.

Rodríguez, Gloria. 2007. "Dinámica del paisaje y efecto de la fragmentación en la diversidad de la vegetación y en la diversidad funcional del suelo de la reserva de la Biosfera de Urdaibai." PhD diss., University of the Basque Country.

Rodríguez, Gloria, Ibone Amezaga, Mikel San Sebastian, Lorena Peña, and Miren Onaindia. 2005. "Efectos de los cambios en los usos del suelo en el paisaje de la Reserva de la Biosfera de Urdaibai." Paper presented at the Biodiversity Loss in Europe Symposium of the Spanish Terrestrial Ecology Association and the European Ecological Federation, March 4–6, Almería, Spain.

Rutledge, Daniel T., and Christopher A. Lepczyk. 2002. "Landscape Change: Patterns, Effects and Implications for Adaptative Management of Wildlife Resources." In *Integrating Landscape Ecology into Natural Resource Management*, edited by Jianguo Liu and William W. Taylor. Cambridge: Cambridge University Press.

Sala, Osvaldo E., F. Stuart Chapin III, Juan J. Armesto, Eric Berlow, Janine Bloomfield, Rodolfo Dirzo, Elisabeth Huber-Sanwald, Laura F. Huenneke, Robert B. Jackson, Ann Kinzig, Rik Leemans, David M. Lodge, Harold A. Mooney, Martin Oesterheld, N. Leroy Poff, Martin T. Sykes, Brian H. Walker, Marilyn Walker, and Diana H. Wall. 2000. "Global Biodiversity Scenarios for the Year 2100." *Science* 287, no. 5459 (March 10): 1770–74.

Santos, Tomás, José L. Tellería, and Roberto Carbonell. 2002. "Bird Conservation in Fragmented Mediterranean Forests of Spain: Effects of Geographical Location, Habitat and Landscape Degradation." *Biological Conservation* 105, no. 1: 113–25.

Santos, Tomás, José L. Tellería, and Emilio Virgós. 1999. "Dispersal of Spanish Juniper *Juniperus Thurifera* by Birds and Mammals in a Fragmented Landscape." *Ecography* 22, no. 2: 193–204.

Saunders, Denis A., Richard J. Hobbs, and Chris R. Margules. 1991. "Biological Consequences of Ecosystem Fragmentation: A Review." *Conservation Biology* 5, no. 1: 18–32.

Sedjo, Roger A., and Daniel Botkin. 1997. "Using Forest Plantations to Spare Natural Forests." *Environment* 39, no. 10: 14–20.

Seto, Karen C., and Michail Fragkias. 2005. "Quantifying Spatiotemporal Patterns of Urban Land-use Change in Four Cities of China with Time Series Landscape Metrics." *Landscape Ecology* 20, no. 7: 871–88.

Stanley, William G., and Florencia Montagnini. 1999. "Biomass and Nutrient Accumulation in Pure and Mixed Plantations of Indigenous Tree Species Grown on Poor Soils in the Humid Tropics of Costa Rica." *Forest Ecology and Management* 113, no. 1: 91–103.

Tellería, José L., and Tomás Santos. 2000. "Fragmentación de hábitats forestales y sus consecuencias." In *Ecosistemas mediterráneos: Análisis funcional*, edited by Regino Zamora and Francisco I. Pugnaire. Madrid: CSIC.

Terradas, Jaume, Raimond Salvador, Jordi Vayreda, and Francisco Lloret. 2004. "Maximal Species Richness: An Empirical Approach for Evaluating Woody Plant Forest Biodiversity." *Forest Ecology and Management* 189, nos. 1–3: 241–49.

Trzcinski, Kurtis, Lenore Fahrig, and Gray Merriam. 1999. "Independent Effects of Forest Cover and Fragmentation on the Distribution of Forest Breeding Birds." *Ecological Applications* 9, no. 2: 586–93.

Turner, I.M., K.S. Chua, J.S.Y. Ong, B.C. Soong, and H.T.W. Tan. 1996. "A Century of Plant Species Loss from an Isolated Fragment of Lowland Tropical Rain Forest." *Conservation Biology* 10, no. 4: 1229–44.

Vitousek, Peter M., Harold A. Mooney, Jane Lubchenco, and Jerry M. Melillo. 1997. "Human Domination of Earth's Ecosystems." *Science* 277, no. 5325 (July 25): 494–99.

Watson, James E.M., Robert J. Whittaker, and Terence P. Dawson. 2004. "Avifaunal Responses to Habitat Fragmentation in the Threatened Littoral Forests of South-eastern Madagascar." *Journal of Biogeography* 31, no. 11: 1791–1807.

Watson, James E. M., Robert J. Whittaker, and David Freudenberger. 2005. "Bird Community Responses to Habitat Fragmentation: How Consistent Are They across Landscapes?" *Journal of Biogeography* 32, no. 8: 1353–1370.

Weathers, Kathleen C., Mary L. Cadenasso, and Steward T. A. Pickett. 2001. "Forest Edges as Nutrient and Pollutant Concentrators: Potential Synergisms between Fragmentation, Forest Canopies and the Atmosphere." *Conservation Biology* 15, no. 6: 1506–14.

Whigham, Dennis F. 2004. "Ecology of Woodland Herbs in Temperate Deciduous Forests." *Annual Review of Ecology, Evolution and Systematics* 35, no. 1: 583–621.

Whitcomb, R. F., C. S. Robbins, J. F. Lynch, B. L. Whitcomb, M. K. Klimkiewicz, and D. Bystrak. 1981. "Effects of Forest Fragmentation on Avifauna of Eastern Deciduous Forests." In *Forest Island Dynamics in Man-dominated Ecosystems*, edited by Robert L. Burguess and David M. Sharpe. New York: Springer.

Wilcove, David S., Charles H. McLellan, and Andrew P. Dobson. 1986. "Habitat Fragmentation in the Temperate Zone." In *Conservation Biology: The Science of Scarcity and Diversity*, edited by Michael E. Soulé. Sunderland, MA: Sinauer.

Yáñez, Marco A., Frida Veral, Javier A. Simonetti, and Audrey Grez. 1999. "Small Mammals of Forest Islands of the Beni Biological Station, Bolivia." *Mastozoologia Neotropical* 6, no. 2: 135–38.

7

Critical Theories of Sustainable Development

Eguzki Urteaga
The University of the Basque Country (UPV/EHU)

Throughout the history of economic thought, critical analytical traditions have existed that have brought to light the ecological damage occasioned by capitalism (Martinez-Alier 1997). The history of debates about nature is marked by a series of problems posed by a "predatory economy" questioned by both classical liberals and socialists: the forest question, at the forefront of attention in the eighteenth century, preceded the coal question, which William Stanley Jevons was interested in during the nineteenth century. These issues stand at some remove from the problem of sustainable development. After having disappeared behind the economic and social consequences of the crisis of 1929 and the two world wars, the idea of a destructive economy returned with a certain degree of force starting in the late 1960s. Numerous economists were concerned about the environmental consequences of the post-war economic miracle. At the same time, they took into consideration the specificity of these phenomena, which cannot be reduced to the logic of the market, with the aim of articulating the fundamentals of the economy and the environment and thereby creating an ecological economics. Despite the fact that the creation of a new discipline that would integrate the knowledge contained in the disciplines of economics and ecology continues to be little more than a proposal, it has the potential to lead to two major political options. The first consists of creating public institutions capable of imposing limits on the exploitation of nature, and the second is forcefully expressed in the field of industrial ecology.

Ecological Economics

Ecological economics is a relatively diverse current of thought, the objective of which is to reflect on the conditions for a possible shared evolution among ecosystems and economic systems. If some authors take the ecological dimensions of natural resources into consideration, others insist more strongly on the institutional issues posed by their regulation.

A Distinction between Growth and Development

One way of thinking about economics makes an almost mechanical association between growth and development, to the point of conflating the two. On the other hand, numerous economists, following in the footsteps of Joseph Schumpeter, try to distinguish them. Growth corresponds to a quantitative increase in available goods and services, measured in monetary and physical terms. Development, for its part, is a measure of qualitative improvement in living conditions. If these objectives differ, they are not necessarily linked, since there is no reason for growth to be a synonym of development. Consequently, the relationship between the two is complex and the source of controversy, and the appearance of the concept of sustainable development has only given new impetus to the debate.

Along these lines, if Wilfred Beckerman (1994) denies any interest in the expression "sustainable development," because he believes that it does not contribute anything pertinent to the analysis, Herman E. Daly (1996) and the supporters of ecological economics denounce as meaningless the idea of "sustainable growth," which is a contradiction in terms. Supporters of ecological economics privilege the analysis of sustainable development as the central focus of ecological economics (Costanza 1991). Beyond the different ways of translating this objective, all of these authors share the idea that from now on, the environment will be the limiting factor in economic development, while for a long period it was capital that played that role. Up to the present, this has had the consequence of focusing the economic calculus on the conditions for the reproduction of capital, which are not the same as those that ensure the reproduction of natural and human resources.

The Biophysical Dimensions of Economic Activity

Ecological economists use thermodynamics to develop a critical analysis of wealth production. If the aim is to evaluate its environmental impact, an evaluation on the basis of prices may be sufficient, without resorting to the concept of utility, which comes from the domain of psychology and does not correspond to any objective measure. Likewise, in most cases, the neoclassical theory of production does not reflect biophysical reality. The constructive logic of this treatment of production is based on the properties of the mathematical functions used and not on their aptitude for representing reality. At best, a rule for the distribution of rent across different social categories can be found (Robinson 1964). Scholars in this field are reactivating an older tradition of thought. Nicholas Georgescu-Roegen (1975), one of the pioneers of mathematical economics, compared the principles of standard economic theory to those of thermodynamics, a mixture, according to him, of physics and economics. This exercise led him to condemn the neoclassical analytical representation that depicts perpetual movement and describes the process of production from the perspective of viewing it as a world-transforming activity.

Thermodynamics is precisely the physical science interested in processes of transformation as they relate to energy. The first principle of thermodynamics stipulates that energy is conserved, and the second claims that energy degrades when a process of energy transformation takes place. Energy then takes on forms that are increasingly less available and useful to individuals. In order to avoid the trap of the "energy myth," Georgescu-Roegen (1975) insisted on the need to extend these properties to matter, to the point of designating material entropy as the fourth law of thermodynamics. To return to one of these examples, the particles of rubber left on the asphalt by car tires have not disappeared, but they have become useless for society. The recovery and reuse of this material that has been dispersed into the natural world demand an undetermined amount of time and a considerable amount of energy. As a result, an endowment of energy is not sufficient to be able to produce goods and services. Economic activity is based on the exploitation of two "low-entropy" sources, the material and energetic resources thanks to which it produces goods and services, as well as "high-entropy" elements in the form of waste. The problems of natural resource depletion and pollution originate in the entropic nature of the economic process. However, the law of entropy does not tell us the speed at which and the manner in which this energetic and material deterioration comes about. Human activity is situated precisely in the space marked out by this indeterminacy.

A Necessary Complementarity

From this perspective, the scarcity combated by economic activity is rooted in sources of low material and energetic entropy. According to ecological economists, this is an absolute scarcity, associated with the name of Thomas Robert Malthus. This conceptual framework has repercussions for analyzing the conditions for economic growth. Accumulated through the use of matter and energy, technical capital also needs matter and energy to function. As a result, it is inconceivable that this kind of capital could completely replace natural capital. These economists emphasize the idea of complementarity between natural capital and the other productive factors that play a role in the processes of production because they are themselves based on low-entropy initial resources. In this analytical framework, the objective of sustainability is defined as the nondecrease over time of the stock of natural capital (Pearce and Turner 1990, 48), enabling the production of a constant flow of wealth in the form of assets and economic and environmental services, such as the purification and recycling capabilities offered by natural systems. Even if ecological economists are only moderately optimistic about technology, they need to take into account the partial replacement possibilities that will arise as a consequence of improvements in the efficiency with which the material and energy flows that play a role in human activities are transformed (Faucheux and O'Connor 1999). This is why some authors prefer to discuss the maintenance over time of a stock of "critical natural capital," to be understood as a number of elements provided by that natural world that is necessary to future generations. This coercion is sometimes characterized as a condition for strong sustainability.

Bioeconomic Models

The instrumental option privileged by ecological economists to achieve such an objective, without neglecting action on prices, consists of establishing quantitative limits with regard to the resources exploited and the waste expelled in nature. They thereby manifest their skepticism of neoclassical analysis with regard to the internalization of externalities. The concept of externality overlooks the fact that the situations they characterize are abnormal, while the logic of economic activity itself is to transform nature and give rise to interdependent relationships among actors. Another reason for doubt arises from the fact that the procedure of internalizing externalities refers to harm—that is, to pollution that has an impact on economic actors' objective functions. The act of taking damage to the environment into account may come too late because businesspeople are scarcely beginning to perceive a decline in their well-being at the point that some ecological thresholds have been exceeded.

Breaking with the idea of integrating into the economic sphere what was initially outside it (the ecological conditions), on the model of René Passet's scheme of three spheres (1979, 11), ecological economists conceptualize the economy as a subsystem of a global system constituted by the sum total of human activities. This is part of a broader system constituted by the biosphere. In this case, the problem lies in creating nested relationships among the different systems: inserting economic matters into social activities, knowing that these, in their turn, should be integrated into the biosphere. The models to which ecological economists explicitly refer (Costanza and Daly 1992, 40) to define the ecological limits on human activity are those of forest or fishing economies. In these bioeconomic models, a biological resource is considered as a kind of natural capital, the management of which should be optimized over the long term.

One of the objectives to be attained is that of maximum sustainable yield, or the maximum amount of resources that can be exploited in each period without putting their regenerative capacity into question. Another interesting aspect of these areas of activity consists in the fact that, very quickly, they have been forced to reflect on the regulatory institutions in a position to respond to these resources' particular characteristics: free access and a reproduction time that far exceeds the usual horizon of economic calculation. The problem is that these models' results are obtained in a static framework, with a hypothesis of perfect knowledge of the stock of resources available, something that is rarely the case in reality.

Starting in the 1980s, the thinking about natural resources management began to take a new form with the rise of global ecology and the recognition of global environmental problems. The fight against pollution and the management of natural resources had henceforward to be considered on the planetary scale. Despite the great acquisition of knowledge related to the environment, it must be acknowledged that the science of the biosphere is in its early stages. In view of the absence of reliable indicators, it is difficult to give operational content to a global bioeconomy. As a result, the aim is to define the optimal magnitude of human activity on the planet, consequently determining the optimal amount of capital created by human beings and the optimal amount of natural capital to be maintained as such. These limits can only be evaluated with reference to the standard of living and the size of the human population considered appropriate.

In reality, the adoption of such a global and monolithic approach is in no one's interest. In fact, natural capital is not a homogeneous whole, since it is made up of a set of resources and environments that differ from one another in their ecological characteristics, their repertories of actors, and, where applicable, the existing regulatory components. As a consequence, it is useful to define specific rules for the different types of natural capital (Pearce and Turner 1990, 43). These principles should be understood as minimum standards of prudence: (1) the rates of extraction of renewable natural resources should be equal to their rates of regeneration; (2) the rates of waste emission should be equal to the assimilation and recycling capacities of the environments in which that waste is expelled; and (3) the exploitation of nonrenewable natural resources should take place in a rhythm equal to that of their replacement by renewable natural resources. These proposals draw on certain ideas formulated in their time by American conservation economists. Daly (1979, 87) refers, for example, to the work of John Ise (1925). Concerned about the depletion of energy and forest resources, Ise advocated for governments to raise prices through taxes, with the aim of influencing, in the case of renewable resources, the demand for these resources and covering their cost of reproduction and, in the case of nonrenewable resources, covering their cost of replacement by other resources. Other conservationists recommended setting up a permit system for the exploitation of natural resources.

Defining Environmental Standards

From this same institutional perspective, Daly proposes the establishment of various bodies to regulate world population, wealth inequality, and the management of natural resources. These proposals resemble the recommendations made by J.H. Dales (1968) with regard to the establishment of a system of transferable quotas to regulate the pollution of large Canadian lakes. In this case, the government initially determines a global quantity of pollutants that is not to be exceeded and subsequently has the right to distribute those pollutants among the economic actors concerned under certain conditions and in the form of permits with expiration dates. Since population and economic activities increase, if the supply of permits is stable, prices will go up, thereby motivating businesspeople to implement techniques to reduce pollution. Dales insistently observes that such a mechanism is an administrative tool that has little to do with a market. It is a planning process associated with a system of quota exchanges that enables firms to adapt as well as possible to this policy. The government does not establish rights equivalent to property rights over land, but it does issue a provisional authorization to expel pollutants. It acts as a broker and a regulator: the public authorities record all transactions and, if they wish, influence permit prices by buying unexpired rights or selling rights kept in reserve. The public authorities have the right to reevaluate the overall amount of pollution authorized at the end of each period during which the rights are valid.

Other supporters of ecological economics (Faucheux and O'Connor 1999) have returned to the concept of "minimal protection standards" for ecological systems, standards that should be the object of negotiations among the different parties involved. In every case, it is a matter of establishing "regulatory management in a framework of coercion," or setting limits to the exploitation of natural resources, defining the most equitable possible conditions for distributing this coercion within society, and specifying the institutions that will enable economic actors to make optimal decisions. In reality, this approach conflicts with the socio-epistemological characteristics of the context in which the negotiations aimed at defining the coercive measures take place.

Environmental Accords

Environmental problems are characterized by their location in "controversial" universes (Hourcade, Salles, and Théry 1992) that make it difficult to define a set of management standards. Uncertainty reigns at all levels of expert opinion and decision making. Even if sufficient scientific knowledge is available about the significance of the challenges posed, fundamental questions remain with respect to the causes, consequences, and responsibilities such that uncertainty is appropriate to invoke in this regard. Moreover, environmental damage is not directly perceived by actors, and certain interests concerned are absent from the negotiations or have contradictory spokespeople. Some economic actors, who approach the problem using an inverse-risk strategy (Roqueplo 1988), try to use science strategically to impact the negotiations. The problem is thus framed by a set of actors who jumble together scientific controversies, industrial interests, political logics, and media effects.

Since it is impossible to have perfect knowledge of the problem posed in order to implement a policy capable of responding to it, the decision-making process little by little becomes autonomous from the scientific controversy. An environmental accord ends up being elaborated, constituting an agreement on the diagnosis of the problem and on the objectives to be attained, as well as institutional solutions or techniques to be implemented. These are normative frameworks created over the course of history and of the negotiating process, frameworks put forward as proposals that enable the actors to make decisions. Although a pronounced tendency to refer to property-rights systems can be observed, there are no general rules in this area, and different negotiations are conducted for matters concerning different natural resources or environments. Some environmental accords, such as those concerning certain fishing areas, already have a long history and evolve over time. Others related to climate change or to the erosion of biodiversity are in the drafting phase. As the example of acid rain demonstrates, nothing guarantees that the response made by an environmental accord will be able to satisfactorily resolve the problem initially posed.

Industrial Ecology

Taking into consideration the biophysical basis of socioeconomic systems, industrial ecology offers another path for reconciling economics and ecology and working toward the advent of a sustainable form of development. The idea of considering the "metabolism" of economic activities, especially agricultural ones, is very old (Fischer-Kowalski 2003). It is also found in Marx's *Capital,* which took inspiration from the first studies conducted by Justus von Liebig on biogeochemical cycles. It also conceptualizes humanity as a true biogeochemical force and draws up eco-energetic calculations. The 1970s, with the oil crisis, saw the spread of analyses of this kind. Allen V. Kneese, Robert U. Ayres, and Ralph C. d'Arge (1970) used the principle of the conservation of matter to model the relationship between the economy and the environment, inaugurating a series of studies on "industrial metabolism" that have not ceased to multiply, in a trend that has occupied a significant place in ecological economics ever since.

At the same time, industrial ecology germinated within the United Nations Environment Programme, which was founded in 1972 with Maurice Strong as director. Fifteen years later, Robert A. Frosch and Nicholas E. Gallopoulos (1989) published a fundamental work on industrial ecology, just after the publication of the Brundtland Report (1987) and before the lead up to the Rio conference (1992). These authors, who worked for General Motors at the time, proposed involving the industrial system in a profound reform of its environmental practices. Once again, the automobile sector showed itself to be on the cutting edge in developing models for the organization of work and production. The spread and institutionalization of industrial ecology took place rapidly in the course of the 1990s, thanks especially to the importance of private consultants. Simultaneously, the business world has come together to publicize its interpretation of sustainable development. Industrial ecology is an element of this conception. This perspective is addressed to the countries of the South (Erkman and Ramaswamy 2003), which find themselves facing heavy pressures in terms of the availability of natural and financial resources, and are inspired by Western models of production and consumption, with their advantages and disadvantages.

Copying Nature

Industrial ecology presents itself as an approach concerned with giving operational content to the idea of sustainable development (Adoue and Ansart 2003), making it possible to escape the sterile debate in which economics and ecology are on opposite sides. The same is true of the members of B4E (Business for the Environment), which calls for ecology and not ideology. It looks to the objective dimension of ecological science in order to rethink the processes of production and consumption. This idea of considering the human and natural economies in a unified way has a certain history. From its origin, thermodynamics has always been a science constructed from the starting point of general reflections that encompass living and technological systems. Beyond the observable differences between these two types of objects, thermodynamics urges a conceptualization of machines as living organisms and vice versa. This same circulation of metaphors and models has accompanied the development of ecosystem ecology since World War II. A spaceship constructed for a long voyage is a good example of an ecosystem (Odum 1971, 64). As a result, it is unsurprising that this line of thought "considers the industrial system as a particular case of an ecosystem" (Erkman 1998, 9).

An additional step has been taken in this approach to systems analysis due to the evolutionary perspective sketched out by some industrial ecologists. If life is "a lasting success," this has not always been the case: "Life has existed on the earth for 3.5 billion years. It did not achieve a sustainable balance . . . until about half that time—nearly 2 billion years—had already passed" (Ayres

1989, 364). These authors retrace the evolution toward the sustainability of life through several major stages. Originally, a type-one ecosystem developed in a low-coercion environment, in which it could choose its resources and expel its waste without cause for concern. Under evolutionary pressure, a type-two ecosystem, more efficient than the previous one, made its appearance, with internal flows of matter and energy more significant than its external ones. Nevertheless, the need to take in some elements from its environment and expel others meant that, after a certain point, the system came up against the limits of its development. Evolution led to the appearance of a type-three ecosystem, the elements of which are continually recycled, because energy continues to be the only external input into the system. The industrial system should follow this same dialectical evolution: "today's industrial ecosystem, based on fossil fuels, resembles the first stages of biological evolution, when the most primitive organisms procured their energy from a stock of organic molecules accumulated during the prebiotic period" (Erkman 1998, 37–38).

An Ecological Approach to Engineering

An ecosystem is a complex ecological system that evolves thanks to the interaction of two components: a biotope and a biocenosis. The *biotope* is the set of abiotic physical and chemical factors that constitute the environment for life within which animal and vegetable species develop. The *biocenosis* is the totality of species associated with an environment for life, developing a network of interdependent relationships through, among other aspects, a food chain. One way to describe this system is by translating these components and their relationships into amounts of energy. By measuring the energy flows that enter and leave the system, it is possible to describe the system's metabolism and the efficiency of the energy transformations that are the basis for the ecosystem's food chain. This is precisely the principle of ecoenergetics. Industrial ecology takes a similar approach to studying the industrial metabolism of socioeconomic systems. Once this is known, four objectives are formulated: (1) optimizing the use of energy and raw materials; (2) minimizing the emission of pollutants and closing the flows that circulate within productive systems; (3) dematerializing economic activities; and (4) reducing dependence on nonrenewable energy sources.

Beyond the study of industrial metabolism, the aim is to construct a type-three ecosystem, or at least get as close to one as possible. The best illustration of this objective is the Kalundborg industrial symbiosis (in reference to the Kalundborg Eco-industrial Park in Denmark), where a refinery a) uses the waste heat from a power plant and b) sells the sulfur extracted from the petroleum it processes to a chemical plant. The refinery also offers calcium sulfate to a producer of wall plaques c) to replace the plaster it usually buys. The excess steam from the power plant likewise d) heats the water used by a fish farm, as well as e) by greenhouses and residences. In the same way that organisms in an ecosystem feed on the waste and remains of other species, the byproducts and waste from one firm serve as raw materials for other firms' products. This set of exchanges of energy and materials linking the leading firms in the Kalundborg industrial zone makes it possible to save resources and produce less ultimate waste. Instead of focusing on the end product, the aim is to avoid producing and emitting new waste.

The Green Invisible Hand

Industrial ecologists so often cite the example of Kalundborg because it can serve as the launching pad for a political message. Commentators on this experience emphasize that these characteristics only entered general awareness in 1989, after thirty years of existence. In other words, this industrial organization is believed to have started spontaneously. If self-organization is one of the properties of an ecosystem, it is also symbolized by the invisible hand. Consequently, industrial ecology is situated in the tradition of the internalization of externalities, leaving the actors to spontaneously negotiate the management of the environment among themselves, on the basis of the cost-benefit calculations they make. Suren Erkman (1998, 26) insists that the exchanges of byproducts and waste that unspool among the different firms in the Kalundborg symbiosis obey the laws of the market.

These entities understand how to create a "win-win" situation, by reducing the environmental impact of their production processes and making money. This doubly beneficial strategy is sometimes called the "Porter hypothesis," which states that strict environmental regulations can induce efficiency and encourage innovations that help improve commercial competitiveness. In all cases, the market is called on to play a role in regulating sustainable development. According to Stephan Schmidheiny (1992, 14), "The cornerstone of sustainable development is a system of open, competitive markets in which prices are made to reflect the costs of environmental as well as other resources."

Management Standards for Self-Regulation

The literature of environmental management puts the emphasis on the manager's role in evolutionary changes. The manager is a providential figure who knows, decides, and convinces others about the changes in orientation to be made. A discourse is also generated relative to the state's role, since it determines the institutional context within which firms act. The environmental policy called for by industrial ecologists is one that privileges financial incentives and reviews standards: "regulations must become more flexible so as not to unduly hinder recycling and other strategies for waste minimization"

(Frosch and Gallopoulos 1989, 151). A set of technical and management standards should certify these eco-efficient practices and make it possible to send signals of quality to consumers and shareholders. Labels are designed to inform the first group, and environmental reports and other management standards the second.

Today, widely disparate practices exist among firms committed to these reporting activities. Several bodies propose different lists of indicators, among which firms choose as they see fit. The quality of these reports and the ways they are validated by rating experts and auditing firms are also highly variable: some refer to their data-collection practices and calculation methods, while others focus on the environmental impact. In the majority of cases, the investigations concern procedures. In this case, these are voluntary commitments by firms, in that they themselves define the environmental objectives they want to attain, taking into account that the establishment of standards testifies to the implementation of appropriate procedures to that end.

The firms' discourse is about a change in their attitude toward environmental issues. The problem of sustainable development is translated at the level of the firm into the problems of social responsibility and responsible investment, which aim to conciliate business activities with the values of society as a whole, responding to the expectations of stakeholders who are not limited to the firm's shareholders and customers.

Self-regulation is considered much more effective than public coercion, since it entails voluntary adhesion by firms and ensures the transparency of the procedures implemented. It is above all a matter of implementing new environmental policies that take into account the interests of the firms responsible. This aspiration is based on a distorted analysis of the policies implemented since the 1970s.

Conclusion

Instead of trying to integrate the environment into the economic sphere, ecological economics aims to define the conditions for inserting economic activity into the biosphere, through a set of socio-environmental standards. For that purpose, however, it is not sufficient to get an approximate idea of some economic and environmental data. The environmental accords that are drawn up confront scientific unknowns and a lack of political will. An additional difficulty lies in the absence of separation between science and politics in setting these standards. On the contrary, the interaction between these two worlds is notable and governs the trajectory of the negotiations, with no certainty as to whether they will respond to the proposed problems. Other difficulties relate to establishing the list of accords necessary in order to take critical natural capital into consideration and to ensure that the different regulatory modes that address phenomena that interact with one another are themselves coherent.

Beyond these practical problems, it is appropriate to emphasize the ambiguity in this current of thought, characterized by a great diversity of viewpoints and political positions. Even if ecological economics opposes the reductionism of neoclassical theory, it does refer to economic rationality. Likewise, reference to bioeconomic models is not specific to ecological economics. The use of the concept of natural capital leads to a highly economistic and instrumental vision of nature. Likewise, if ecological economics poses the question of the transmission of our resource stock, that should lead us to reflect on the concept of natural patrimony, which offers an alternative to standard economic reference points. It equally leads to new issues, because patrimonialization is a complex phenomenon that brings into play the formation of human communities around certain objects for the purpose of inscribing themselves in a particular way in time (Barthélemy, Nieddu, and Vivien 2004).

The ambiguity is even greater when industrial ecology is considered. This approach can be recovered by the business world that aims to incarnate sustainable development in the future. The road to a form of eco-capitalism is open. Nevertheless, the problem of the establishment of global limits remains relevant, since a decrease in individual consumption of energy and materials may be balanced out by a global increase in the consumption of products. For example, new vehicles consume less gasoline, despite an increase in automobile traffic. The issue also arises of the alternate sources of authority based on voluntary commitments and partnerships with NGOs, given that critical attitudes toward capitalism have softened over time.

References

Adoue, Cyril, and Arnaud Ansart. 2003. "L'essor de l'écologie industrielle: Une avancée vers le développement durable." *Futuribles* 291 (November): 51–67.

Ayres, Robert U. 1989. "Industrial Metabolism and Global Change." *International Social Science Journal* 121 (August): 363–73.

Barthélemy, Denis, Martino Nieddu, and Franck-Dominique Vivien. 2004. "Externalités ou production de patrimoines? Les enseignements de travaux récents sur l'agriculture et l'environnement. " *Géographie, Economie et Société* 6, no. 3: 331–52.

Beckerman, Wilfred. 1994. "Sustainable Development: Is It a Useful Concept?" *Environmental Values* 3, no. 3: 191–209.

Costanza, Robert, ed. 1991. *Ecological Economics: The Science and Management of Sustainability*. Technical editor, Lisa Wainger. New York: Columbia University Press.

Costanza, Robert, and Herman E. Daly. 1992. "Natural Capital and Sustainable Development." *Conservation Biology* 6, no. 1: 37–46.

Dales, J. H. 1968. *Pollution, Property and Prices*. Toronto: Toronto University Press.

Daly, Herman E. 1996. *Beyond Growth: The Economics of Sustainable Development*. Boston: Beacon Press.

———. 1979. "Entropy, Growth, and the Political Economy of Scarcity." In *Scarcity and Growth Reconsidered*, edited by V. Kerry Smith. Baltimore: Johns Hopkins University Press.

Erkman, Suren. 1998. *Vers une écologie industrielle*. Paris: C.L. Mayer-la Librairie.

Erkman, Suren, and Ramesh Ramaswamy. 2003. *Applied Industrial Ecology: A New Platform for Planning Sustainable Societies*. Bangalore: Aicra Publishers.

Faucheux, Sylvie, and Martin O'Connor. 1999. "Un concept controversé: le capital naturel." *Cahier du C3ED* 99-01. Guyancourt, France: Centre d'Economie et d'Ethique pour l'Environnement et le Développement, Université de Versailles Saint-Quentin-en-Yvelines.

Fischer-Kowalski, Marina. 2003. "On the History of Industrial Metabolism." In *Perspectives on Industrial Ecology*, edited by Dominique Bourg and Suren Erkman. Foreword by President Jacques Chirac. Sheffield: Greenleaf.

Frosch, Robert A., and Nicholas E. Gallopoulos. 1989. "Strategies for Manufacturing." *Scientific American*, special issue, "Managing Planet Earth," 261, no. 3: 144–52.

Georgescu-Roegen, Nicholas. 1975. "Energy and Economic Myths." *The Southern Economic Journal* 41, no. 3: 347–81.

Hourcade, Jean-Charles, Jean-Michel Salles, and Daniel Théry. 1992. "Ecological Economics and Scientific Controversies: Lessons from Some Recent Policy Making in the EEC." *Ecological Economics* 6, no. 3: 211–33.

Ise, John. 1925. "The Theory of Value as Applied to Natural Resources." *The American Economic Review* 15, no. 2: 284–91.

Kneese, Allen V., Robert U. Ayres, and Ralph C. d'Arge. 1970. *Economics and the Environment: A Materials Balance Approach*. Baltimore: Johns Hopkins University Press.

Martinez-Alier, Joan. 1997. "Some Issues in Agrarian and Ecological Economics: In Memory of Georgescu-Roegan." *Ecological Economics* 22, no. 3: 225–38.

Odum, Eugene P. 1971. *Fundamentals of Ecology*. Philadelphia: W.B. Saunders.

Passet, René. 1979. *L'économique et le vivant*. Paris: Payot.

Pearce, David W., and R. Kerry Turner. 1990. *Economics of Natural Resources and the Environment*. London: Harvester Wheatssheaf.

Robinson, Joan. 1964. "Kalecki and Keynes." In *Collected Economic Papers*. Volume 3. Oxford: Basil Blackwell.

Roqueplo, Philippe. 1988. *Pluies acides: Menaces pour l'Europe*. Paris: Economica.

Schmidheiny, Stephan. 1992. *Changing Course: A Global Business Perspective on Development and the Environment*. Cambridge, MA: MIT Press.

8

Bases for the Transition toward a Sustainable Economy

ROBERTO BERMEJO, DAVID HOYOS
University of the Basque Country (UPV/EHU)

ENEKO GARMENDIA
Basque Centre For Climate Change (BC3)

In general, pre-capitalist societies demonstrated a high level of equilibrium with the environment. It is interesting, however, that the physiocratic school of economics, which appeared at the dawn of capitalism, sought to construct a sustainable economy in complex societies. In facing the nonsustainability of industrial civilization, solutions from the system of power have focused only on technological change. Although in order to move toward sustainability we must replace the majority of dominant technologies with others that are environmentally respectful, this option is insufficient among other reasons because of the Jevons Paradox; the notion that technological progress that increases the efficiency with which a resource is used tends to increase (rather than decrease) the rate of consumption of that resource (Polimeni et al. 2008). Structural transformations are necessary and, by looking to nature, we can infer the principles of a sustainable economy. The objective of this chapter is to specifically understand the operation of natural systems in order to establish these principles.

Sustainability and Economy

Up until the seventeenth century, an organicist and religious view of the world predominated, according to what was granted life. Human beings could not substantially alter the creation of wealth, but could act within the natural order to maximize physical production. Although for the duration of this long period the norm was as described, there were also notable exceptions that in many cases caused collapses.[1] However, there were also societies that

1. On Easter Island, the competition between the Polynesian colonists to construct giant effigies led to the destruction of the local ecosystem, and the population was reduced from ten thousand to just a few hundred. Those remaining were left in a very precarious state with permanent war between them. The Anazari constructed large settlements with multi-floored housing. When the colonists arrived, the settlements were empty because essential natural resources had been exhausted.

corrected their nonsustainable dynamic and they survived. We also find cases that included a spectacular correction of nonsustainable behaviors.[2]

Physiocrats departed from the religious concept of economics but maintained the organicist theory and concern for the physical–natural foundation it is based on. They believed that human beings were capable of increasing and willfully controlling production through work, with the help of science. Thinkers in the seventeenth and eighteenth centuries stayed within the land–labor binomial, with variation only in the influence given to each of the two elements. Thus nature placed limits on work, and only by respecting that principle could they guarantee unlimited reproduction of economic activity. Physiocrats were concerned about the utilitarian nature of production and reasoned in monetary terms. Money was their only means of measurement and that led to criticism from classical economists (Naredo 1987, 104). Thus, at the end of the eighteenth century, with Adam Smith the foundations of the prevailing classical economic system were established, based on monetary reduction.

However, classical economists considered an economy with perpetual economic growth to be impossible. For David Ricardo, a population explosion would cease once the maximum production of food was reached; the economy would then stagnate. He proposed that this structural trend could only be overcome through intervention by the state, thus giving rise to the neo-Ricardian or Keynesian school. Marx would subsequently criticize capitalism's exploitation of the earth. However, John Stuart Mill proposed a sustainable economic vision by advocating a steady state economy, or one that evolved and was improved but that did not grow physically. In the twentieth century, the prestigious mathematician and economist Rumano Nicholas Georgescu-Roegen, considered to be the father of ecological economics (which he called *bioeconomics*), established the foundations for a sustainable economy. According to him, the very mechanism of orthodox economics is what makes it incapable of providing adequate answers for a reality it is unable to explain. Georgescu-Roegen's belief was that economic growth is impossible because it leads to a depletion of energy sources, given that it is dispersed through use based on the second law of thermodynamics, which shows that usable energy is dispersed and cannot be used again. On this basis, he became a defender of evolution and transdisciplinarity because the bioeconomy could not be constructed from the economy (Carpintero 2006,

2. The Polynesian colonists who occupied the South Pacific island of Tikopia 3000 years BP were destroying the local ecosystem in order to feed a population that had been expanding over 15,000 years. Faced with the nonsustainability of the process, they developed an agricultural production system similar to what is now called *permaculture*, creating a highly productive ecosystem of trees and a wide variety of plants. Furthermore, they stabilized the population (Gowdy 2006).

103–5). Herman E. Daly (1974), a follower of Georgescu-Roegen and the modern leader of ecological economics, reclaims and deepens Mill's vision of the steady state. His principles include population constancy and constructed capital (infrastructures, buildings, machinery, etc.), as well as reducing utilization of nonrenewable energy and materials as much as possible.

Alfred Marshall, the great founder of the neoclassical school, considered that biology offered a paradigm that was more relevant for economics. Wassily Leontief also asked that other sciences help to remove dominant economics from the "splendid isolation in which it is found" (Hall and Klitgaard 2006), a vision also defended by the Nobel Prize winners Trygve Haavelmo and Jan Tinbergen. Likewise, the manifest from a noteworthy group of several scientists (led by another Nobel Prize winner, Kenneth Arrow) also warns of the rapid destruction of the natural resources base: "This resource base is finite," although it describes the possibility that the economy and population will grow "at least for some period of time," but to do so, "we need to ensure that the ecological systems on which our economies depend are resilient" (Arrow et al. 1995). Therefore, as Janine Benyus reminds us, "we don't need to invent a sustainable world—that's been done already" (cited in Hawken, Lovins, and Lovins 1999, 73).

Natural Systems as Complex Adaptive Systems

Although linear thinking is one of the distinctive features of the dominant paradigm, it is paradoxical that complex systems are proliferate in the natural world. According to Robert Costanza et al. (1993, 545), a complex system is characterized by strong (usually nonlinear) interactions between parts, complex feedback loops that make it difficult to distinguish cause from effect, and significant time and space lags, discontinuities, thresholds, and limits, all of which result in scientists' inability to simply add up or aggregate small-scale behavior to arrive at large-scale results. Everything is connected, which therefore ascertains that when an element suffers a direct impact, the remaining elements are affected. And a small disturbance can trigger changes that are disproportionately large. When these systems have the capacity to adapt to changes in the environment, they are defined as *complex adaptive systems* (CAS) (Holling 1978; Gunderson 1999; Gunderson and Holling 2002). Through nonlinear interactions between their components, CASs organize hierarchically into structures that determine and are reinforced by flows of individuals, materials, energy, and information. CASs are open and dissipative. They need exterior contributions because they are continually sending materials and, above all, energy into their environment. Natural CASs do not lose materials other than those dissipated, and said flow tends to decrease as they mature. The following are examples of a CAS: a cell, the brain, the immune system, an organism, social insect colonies, ecosystems, the biosphere, an economy, and a society (Matutinovic 2002, 422; Nielsen 2007, 6, 9, 16). Natural CASs are characterized by their stability.[3] Ecosystems are the first basic unit capable of being autonomous. They also constitute the lowest level capable of closing flows of materials and the best structure for capturing and utilizing energy (Hall and Klitgaard 2006).

Ecosystems have many compartments, with many connections. In fact, in nature, "everything is linked to everything" (Nielsen 2007, 6). The flows are of nutrients, water, toxins, energy, individuals, and information. Thus, the ecosystems integrate all biotic parts and some abiotic ones. They control flows of materials and energy and have defined limits. The species or systems that form them are highly interdependent and are strongly tied to feedback and coevolution with the environment. As a natural system matures, its stabilizing mechanisms become more sophisticated. The rates of change for species are very different and are faster than the rates of change for ecosystems. Therefore, the different rates of activity for the components of an ecosystem act as its balancing mechanism. In turn, an ecosystem forms part of another larger one that gives it stability and exercises its hierarchy over the systems that comprise it. In nature, the hierarchy is determined by each species' level of contribution to the continuance of the ecosystems. However, an ecosystem is formed by subsystems that develop faster than it does and, therefore, they are the origin of the ecosystem's adaptive changes.

Therefore, successful development of a system is determined by two complementary elements: growing differentiation and diversity and its integration into an increasingly more complex hierarchical structure. These characteristics provide high stability because they make it possible for the system to maintain functionality in the face of disruptive processes (droughts, fires, plagues, etc.) (Levin 2000, 14, 15).

The Principles of Sustainability for Socioeconomic Systems

Life on Earth is preserved and developed because ecosystems carry out three fundamental abiotic functions: closing material flows; utilizing solar energy; and intervening on the abiotic or inert environment to maintain its physical–chemical properties within an interval suitable for preservation of life. To guarantee abiotic functions, ecosystems show some typical characteristics: evolution, diversity, hierarchical structures, self-sufficiency, decentralization, competency, and, last and foremost, cooperation. Finally, nature has "eco-

3. It's important to distinguish this concept from the theory of general equilibrium proposed by neoclassical economics and upon which the orthodox branch of environmental economics has been founded (see Faber, Manstetten, and Proops 1996).

nomic" structures in that it produces materials through photosynthesis and consumption occurs along the food chain, which produces an exchange of "assets." Therefore, the human economy can be nothing more than a subsystem of this general economy of materials and energy, which is the nonliving component (abiotic) of ecology because a subsystem cannot transgress the rules of the system it belongs to.

Thus the human economy will only be sustainable when it imitates nature and respects the limits established therein. The concept of biomimicry involves converting abiotic behaviors into abiotic principles. And biotic behaviors give rise to biotic principles that, in essence, are summarized in the ecosystemic behavior: evolution, diversity, self-sufficiency, decentralization, natural hierarchy, and the predominance of cooperation over competition.

Abiotic Principles

Of the large number of simple organic elements and compounds in nature, only some are essential for life, specifically those called *biogenic substances* or nutrients. As ecosystems develop, they tend to increase their levels of cooperation and self-sufficiency, recycling of materials, and the period required for their renewal and accumulation. Waste is converted into the primary source of nourishment, thereby also reducing abiotic contributions from the exterior. Nature tends to maintain old niches and to create new ones, and this proliferation of niches is one of the functional mechanisms that explain the closure of the material cycles.

Contrary to natural systems, socioeconomic systems (SSs) produce enormous and diverse amounts of waste that is deposited into nature. On a planetary scale, 57,643 million tons (mt) of materials (known as direct materials) are used annually, broken down into: 18,024 mt of biomass; 11,602 mt of fossil fuels; and 28,017 mt of minerals (broken down into: 16,728 mt of construction minerals, 4,862 mt of industrial materials, and 6,427 mt of metals). However, to be able to obtain these materials, others must be excavated, using large amounts of oxygen and water along with causing massive soil erosion in areas of industrial agriculture. These materials constitute the hidden flows in the productive processes, and their dimensions are enormous: 27,544,293 mt (www.materialflows.net).

The materials cycle and, ultimately, the life cycle cannot be maintained without intake of a permanent flow of energy that, by demand, must be renewable. Fixed energy is passed to other organisms through food chains in a cascading process and works to produce food and create new individuals, dissipating what remains in the form of low-temperature heat. This flow moves in only one direction and is not reversible because the energy dissipates with use. Therefore, solar energy must flow permanently.

On the contrary, SSs are characterized by their behavior in opposition to mature ecosystems. They almost exclusively use nonrenewable energy resources. Consumption of fossil fuels represents over 80 percent of all global commercial energy and, with the inclusion of nuclear energy, more than 90 percent of energy used is nonsustainable. Added to that, our energy system is highly inefficient. Generation of electricity from fuels does not achieve 50 percent performance, and combustion engine performance is around 20 percent. However, wind, photovoltaic, and solar-thermal power production are growing at annual rates of 30–60 percent.

The biotic (live) environment is acting upon the inert (abiotic) environment so that it can maintain some physical–chemical characteristics that support life. This means that the two environments have coevolved over time in order to form a complex and self-regulated system (Levin 2000, 28). The elements that are physiologically controlled include: "surface temperature, atmospheric composition of reactive gases, including oxygen, and pH or acidity-alkalinity" and preservation of the high level of salinity of the oceans (Margulis 1998, 121, 123). But there are other considerations involved. The composition of gases that make up the atmosphere is not completely stable because some react among themselves (especially oxygen and methane). However, an extraordinarily stable composition has been maintained over time because the biosphere has been emitting gases and absorbing others (especially carbon dioxide [CO_2]) in amounts needed to maintain an equilibrium. The earth has a natural tendency to increase its level of acidity, and the large majority of live organisms can only withstand a certain level of acidity. Due to the action of nitrifying microorganisms, the biotic environment is able to produce an adequate amount of ammonia (NH_3) to prevent the acidification process. Nature also purifies the air and water, detoxifies waste, controls diseases, and creates an ozone layer that provides protection from damaging levels of ultraviolet radiation (Odum and Sarmiento 1998, 72; IETC 2003, 21).

The Gaia theory, designed by James Lovelock and developed by Lynn Margulis, describes harnessing the planetary ecosystem's ability to keep the physical–chemical characteristics of the abiotic environment stable. From this perspective, the biosphere can be defined as "a super ecosystem (but not a super organism, because its development is not genetically controlled . . . it is the series of interacting ecosystems that compose a single huge ecosystem at the Earth's surface [that] behaves as a physiological system in certain limited ways" (Margulis 1998, 120–23). This vision concurs with the prevailing thinking of pre-capitalist societies and from the greater part of the eighteenth century, where nature was seen as a self-organized and self-regulated system.

Our civilization is acting in opposition to the Gaian dynamic. The human species is altering biogeochemical cycles. Nitrous and sulfur oxides are acidifying the soil and water and the increase in CO_2 concentrations in the ocean is becoming the primary factor. However, we can regenerate degraded or destroyed natural spaces, transforming the current technosphere into one that is in harmony with nature, following the example of rural communities that increased biodiversity in their environments and improved the quality of their land.

Biotic Principles

The permanence of abiotic functions can only be guaranteed if biotic communities organize into ecosystems, but their characteristics can only be understood from a systemic perspective (integrated or interrelated).

The Darwinian point of view presents evolution that is the result of the accumulation of numerous small changes due to competition between species. On the contrary, the emerging point of view shows evolution based on the occurrence of general leaps. Thus ecosystems grow until reaching maturity and then they evolve according to an unstable equilibrium process, and this evolution is the result of multiple processes that interact on different scales (Rammel, Stagl, and Wilfing 2007). On one hand, the groups of organisms coevolve. On the other, biotic evolution depends on the evolution of the abiotic environment. Biophysical processes develop on enormously varying scales of time and space.

SSs also evolve, but almost always do so based on subjective criteria, such as consumer preferences, market mechanisms, politics, or the interests of large transnational companies (Nielsen 2007, 8). Evolution according to subjective criteria implies unlimited economic product growth and uncontrolled population growth. Based on the industrialization process principle, the human species begins to grow meteorically. Along with that, the market economic system must also grow without limit. While ecosystems tend to reach a state of maturity, the current economic system is not capable of reaching maturity, "as the time scales of evolution of new techniques and technologies are shorter than the time it takes to establish a stabilized situation" (Nielsen 2007, 15). The market economic system needs to grow because it is driven by three factors: capitalization of investments, easing tensions caused by a polarization of wealth, and defraying costs of increasingly complex social structures (health, education, social security, and so on) (Matutinovic 2006, 3). On the contrary, in the last few decades, noted economists have been defending a steady state economy and even a decrease in "developed" countries (Daly 1974, 1992; Latouche 2008).

Along with evolution, the idea of cooperation over competition is one of the characteristic features of biotic communities. There are six major types of interactions between two or more species: competition, predation, parasitism, commensalism, cooperation, and mutualism. With competition, the result of the interaction is negative for the species involved. Predation is positive for the predator and negative for the prey. Parasitism is negative for the host and positive for the parasite. Commensalism is a simple form of positive interaction where one species benefits and the other appears unaffected. Cooperation is when species mutually benefit, although the benefit is not critical for any of them. Mutualism is a relationship that is vital or extremely necessary for the survival of the species involved.

The Darwinian theory of "survival of the fittest" is a simplification of reality because it exclusively focuses on negative interactions (competition, predation, and parasitism) (Levin 2000, 20). Meanwhile, competition plays a lesser role in evolution and, on the contrary, mutualism explains the constant growth of biodiversity. If competition were dominant, a strong force would be produced, causing a reduction of species and an evolution toward more simplified and specialized ecosystems. The fact that biodiversity is increasing indicates that the selection mechanism is "weak at best" (Levin 2000, 185) or imperfect. For many authors, "the maintenance of variability and diversity is believed to be mainly a passive process due to imperfection of diversity-reducing principles like selection and competition" (Rammel and Staudinger 2002, 303).

The idea of competition over cooperation is one of the primary causes of the nonsustainability of SSs. For orthodox economics, "free competition" is the only mechanism capable of achieving an efficient economy and efficiency means "cost minimization, factor endowments and the exclusion of 'weak' performers" (Rammel and van den Bergh 2003, 128). Furthermore, this concept rests on the premise that efficient action by each agent gives rise to an improved economic system (a linear approach, not systemic). On the contrary, it overlooks that the pursuit of improved efficiency by companies causes systemic inefficiencies. The most employed mechanism is a systemic reduction of costs and, in particular, of staff, which results in high rates of structural unemployment.

Frequently, whatever the companies have saved they then have to pay to society in the form of unemployment subsidies. In the event that competition causes a crisis for major companies, governments frequently invest in bailing them out. This policy has reached its pinnacle in the recurring crises that the market economy system has suffered. In the 2008–2009 global economic crisis, governments gave billions of dollars to banks and major companies, which constitutes a flagrant infringement of free market principles. In summary, the market economic system is highly inefficient.

However, we must also analyze dynamic efficiency or the capacity to maintain the identity of a system when facing environmental impacts and the capacity to avoid processes that are self-destructive due to their intrinsic non-sustainability. The fact that historically periodic crises have occurred shows that this is a structural feature of the system, as Marx demonstrated. On the contrary, dynamic efficiency implies maintaining "the greatest potential for future development of a system." And this leads to a high level of diversity, high cooperation, and a high level of systemic competence understood to be the "implied capability to understand the superiority of the systemic viewpoint without giving up individual interest" (Schütz 1999, 26–28).

Diversity has been naturally increasing over some 4 billion years, despite the occurrence of five mass extinctions. Biodiversity reinforces the stability of ecosystems. High biodiversity results in a high level of redundancy, which increases stability.

In addition, although the economic system continually creates new technologies and products, the system trends toward uniformity. Globalization implies the creation of global productive structures and universal technological systems. The resulting head-on competition becomes the only mechanism for selection, leading to growing specialization and uniformity, "and therefore a reduction of diversity, ethnic diversity, cultural diversity, language diversity, and diversity of tastes, preferences, and values" (Joung et al. 2006, 311). This fact and the erosion of social cohesion deserve special attention because they are the supports that societies are built upon in the long term (Schütz 1999, 25, 26; Rihani 2002, 109). The greater part of innovation is incremental and "dominated by economic efficiency, short-term optimization and increasing returns" (Rammel and Staudinger 2004, 16).

Thus the free trade dynamic reduces the capacity to take advantage of the human and natural resources in the local environment and their innovative potential. And as diversity decreases, the potential for responding to disturbances and creating innovative combinations also decreases. Therefore, many authors (Folke et al. 2005; Joung et al. 2006) claim there is the need for polycentric government systems on an institutional level, which could be known as multilevel governing (see Paavola 2007).

Along with increased differentiation and diversity, successful development of a system is determined by its integration into an increasingly complex hierarchical structure (Schütz 1999, 108–9). The natural hierarchy is scalar, self-organizing, embedded, inclusive, and complex. Each natural subsystem is subject to the hierarchy of the system of which it is a part and subject to the parts that form it. However, this subordination between levels is always incomplete and each level has its own behavioral rules and its own relationships.

The natural hierarchy is embedded because it is determined by its dependency on the system that contains it. The hierarchy of the highest levels is because it establishes limiting conditions for behavior at the lower levels, although the natural hierarchy is not exclusive. In existing ecosystems, there is a hierarchy of species because the functions they perform have an unequal importance. However, this hierarchy changes with various circumstances. Therefore, two complementary aspects determine the successful development of an ecosystem: its increasing differentiation and diversity, and its integration into an increasingly complex hierarchical structure.

SSs are also organized hierarchically, but "there are huge differences in the character of the hierarchies" between them and the natural systems (Nielsen 2007, 14). The primary difference is that, while natural hierarchies are predominately scalar, SSs have a control hierarchy. In SSs, the hierarchies have much fewer levels than in natural systems, they are not self-organizing or embedded, and they are exclusive (Nielsen 2007, 14). The hierarchical structure is not embedded because it is predetermined by the decision-making structure (the level of existing democracy, channels of communication and participation, existence of groups of power apart from any political representation, etc.), not by the structure of the system. It is exclusive because the power is highly centralized, and consequently very few individuals truly exercise control. This hierarchical structure makes SSs especially vulnerable and very prone to collapse.[4]

Finally, as ecosystems evolve, they become more self-sufficient, reducing their dependency on forces out of their control. Ecosystems have natural limits due to changes in environmental conditions or the self-organization of the system itself. Spatial limits favor increased internal efficiency of the system and represent the existence of natural growth limits (Ring 1997, 242).

A globalized market eliminates spatial distinctions and produces an enormous flow of resources from poor countries to rich ones, giving the false sensation that the resources are unlimited and using them efficiently is not a concern. The fact that the resources are becoming depleted and that their extraction has given rise to enormous environmental impacts is unknown to the majority of the consumers of these resources. In fact, when a community uses its own resources and develops its own self-regulating institution, it tends to be concerned about their depletion and about the environmental impact of their extraction and collection.

A decentralized model is the most efficient and democratic way to satisfy a population's vital needs. Decentralization makes it possible to create multiple social systems adapted to the conditions of their environments by developing organizational models and institutions that are adapted to local characteris-

---

4. For more information about the relevance of institutional diversity in the sustainable management of natural resources, see Ostrom (1990, 2009) and Dietz, Ostrom, and Stern (2003).

tics; by developing local technologies or global technologies adapted to utilize local resources; and by accumulating experience in the sustainable use of local resources. Decentralization is the only viable way to achieve a cyclical materials economy because the consequences of extraction, transformation, and consumption are very closely linked. Solar energy, for example, is decentralized and therefore so is its capture. To live on solar energy implies no longer depending on a few countries that hold the majority of fossil fuels that are now running out (Joung et al. 2006, 310).

The idea of self-sufficiency is not new and has been championed from Keynes (1933) to the 1992 Rio Declaration by the United Nations and the European Commission's Aalborg Charter (1994). Movement by societies that are facing an energy emergency (the "transition towns" movement represents a noteworthy example because of its growth) implies a qualitative leap in this direction, motivated by the developing energy crisis.[5]

Conclusions

Physiocrats were very aware that natural resources needed to be preserved. Classical economists reclaimed essential elements of physiocratic principles and thought that natural resources (the earth) and work were the foundations of wealth and would transform resources into assets, although productivity was improved with capital contribution. Thus their theory was macroeconomic (Gowdy 2006; Hall and Klitgaard 2006).

This macroeconomic vision was shunted by the so-called neo-classicist or marginalist revolution that began in the 1970s and was replaced by one centered on microeconomic analysis (the company in its market) by which any limitation on natural resources was banished. That this occured in the 1970s was not coincidental because the economy at that time was based on then-abundant carbon, and with oil just beginning to be introduced, all of its enormous reserves were available. In addition, the microeconomic focus developed in aggregate at a point when the general economic system had reached a "general equilibrium," due to the play of forces acting in a free market. In this context, the new economic driver became "technology," an abstraction that frequently has little to do with the physical realities of production and that is attributed nearly magical properties, such as being able to eliminate any shortage of natural resources (Gowdy 2006).

Placing the neoclassicist economy into the monetary universe was objectively constituted within an autonomous field of knowledge unqualified by other sciences. The hegemony of the neoclassical economy and, ultimately, the belief in perpetual growth, have been based on the existence of enormous

---

5. For more information, see www.postcarboncities.net.

fossil fuel resources that have great calorific power per unit of weight, a wide range of uses, and that have been easily accessible. But these resources have now clearly entered into a depletion process (Bermejo 2008; Kerschner, Bermejo, and Arto 2009). Moreover, exploiting these resources has not resulted in the resolution of very pressing problems, such as the polarization of wealth and hunger.

Therefore, the neoclassical economy must make way for another that will meet the principles of a sustainable economy. To do so will require an epistemological revolution. In this context, transdisciplinary approaches that make an "orchestration" of sciences possible or that integrate multiple disciplines acquire special relevance (Neurath 1973; Max-Neef 2005). This would allow a break from the isolation and hegemony of the neoclassical economy and would subject the analysis of economic behavior to the laws that govern other disciplines. Along these lines, environmentalism must play a key role, considering that it is a "super science," or a unified science.

An epistemological revolution would also involve foregoing the distorted ideas of key concepts such as production, development, wealth, and so forth. *Production* is understood as an increase in monetary wealth and is characterized as entirely positive, independent of the depletion of natural resources that it causes and the resulting environmental and oftentimes social impacts. This is not only applicable to the production of goods, but also to the extraction of natural resources, which is a fallacy. Aligned with this idea, the concept of development is defined as the volume of revenue per capita, and that figure is defined as wealth (Naredo 2006, 66, 67, 177–82). This does not take other types of wealth into consideration, such as full satisfaction of needs, high social cohesion, high levels of democracy, availability of water and healthy food, or the absence of contamination. Neither is the destruction of the environment included in the definition of wealth.

Sustainability requires also that the neoclassical economy suffer another epistemological revolution: to use physical calculations in addition to monetary ones. Use of money is absolutely necessary to organize the economies of complex societies, but monetary value is a veil that covers the biophysical reality. To be able to design a sustainable economy, there must be a departure from the knowledge of resources and rates of consumption (and therefore of depletion) of materials (abiotic and biotic) and energy (fossil fuels and renewable). To construct a solar economy, for example, the potential of various renewable energy sources and the status of the technologies used for capture must be identified. To maintain or improve services that give us ecosystems, we must know their conditions and evolution. Based on them, the physical economy must study the metabolism of our societies through calculations of the flow of materials (energy and nonenergy), as a way to determine strategies to close materials flows and live from solar energy.

Lastly, industrial economies are especially nonsustainable and therefore there is a greater urgency and obligation to transform them. They have far exceeded safe limits and must transform in order to decrease environmental risk. Countries with less industrialized economies must continue their development process by using sustainable models so that they do not exceed a safe limit and so that they can instead reach a level of productive maturity through sustainable means. They can then achieve the same type of sustainable economy as industrialized countries that have gone through a transformation.

References

Arrow, Kenneth, Bert Bolin, Robert Costanza, Partha Dasgupta, Carl Folke, C.S. Holling, Bengt-Owe Jansson, Simon Levin, Karl-Göran Mäler, Charles Perrings, and David Pimentel. 1995. "Economic Growth, Carrying Capacity, and the Environment." *Science* 268 (April 28): 520–21.

Bermejo, Roberto. 2008. *Un futuro sin petroleo: Colapsos y transformaciones socioeconómicas, Economia crítica y Ecologismo social.* Madrid: Catarata.

Carpintero, Oscar. 2006. *La bioeconomía de Georgescu-Roegen.* Madrid: Montesinos.

Costanza, Robert, Lisa Wainger, Carl Folke, and Karl Göran-Mäler. 1993. "Modeling Complex Ecological Systems: Toward an Evolutionary, Dynamic Understanding of People and Nature." *BioScience* 43, no. 8: 545–55.

Daly, Herman E. 1974. "The Economics of the Steady State." *American Economic Review* 64, no. 2 (May): 15–21.

——. 1992. "Allocation, Distribution, and Scale: Towards an Economics that Is Efficient, Just, and Sustainable." *Ecological Economics* 6, no. 3: 185–93.

Dietz, Thomas, Elinor Ostrom, and Paul C. Stern. 2003. "The Struggle to Govern the Commons." *Science* 302, no. 5652 (December 12): 1907–12.

Faber, Malte, Reiner Manstetten, and John Proops. 1996. *Ecological Economics: Concepts and Methods.* Cheltenham, UK: Edward Elgar.

Folke, Carl, Thomas Hahn, Per Olsson, and Jon Norberg. 2005. "Adaptive Governance of Social-Ecological Systems." *Annual Review of Environment and Resources* 30: 441–73.

Gowdy, John M. 2006. "Evolutionary Theory and Economic Policy with Reference to Sustainability." *Journal of Bioeconomics* 8, no. 1: 1–19.

Gunderson, Lance. 1999. "Resilience, Flexibility and Adaptive Management—Antidotes for Spurious Certitude?" *Conservation Ecology* 3, no. 1: 7.

Gunderson, Lance H., and Crawford S. Holling. 2002. *Panarchy: Understanding Transformations in Human and Natural Systems.* Washington, D.C.: Island Press.

Hall, Charles A. S., and Kent A. Klitgaard. 2006. "The Need for a New, Biophysical-based Paradigm in Economics for the Second Half of the Age of Oil." *International Journal of Transdisciplinary Research* 1 no. 1: 4–22.

Hawken, Paul, Amory Lovins, and L. Hunter Lovins. 1999. *Natural Capitalism: Creating the Next*

Industrial Revolution. London: Little, Brown and Company.

Holling, Crawford S., ed. 1978. *Adaptive Environmental Assessment and Management*. New York: John Wiley & Sons.

IETC (International Environmental Technology Center). 2003. *Environmentally Sound Technologies for Sustainable Development*. Paris: PNUMA.

Joung, Oran R., Frans Berkhout, Gilberto C. Gallopin, Marco A. Janssen, Elinor Ostrom, and Sander van der Leeuw. 2006. "The Globalization of Socio-ecological Systems: An Agenda for Scientific Research." *Global Environmental Change* 16, no. 3: 304–16.

Kerschner, Christian, Roberto Bermejo, and Iñaki Arto. 2009. "Petróleo y carbón: Del cenit del petróleo al cenit del carbón." *Ecología Política* 39: 23–36.

Keynes, John Maynard. 1933. "National Self-sufficiency." *Yale Review* 22, no. 4: 755–69.

Latouche, Serge. 2008. *La apuesta por el decrecimiento*. Translated by Patricia Astorga. Barcelona: Icaria. French version: *Le pari de la décroissance*. Paris: Fayard, 2006.

Levin, Simon A. 2000. *Fragile Dominion: Complexity and the Commons*. Cambridge, MA: Perseus.

Margulis, Lynn. 1998. *The Symbiotic Planet: A New Look at Evolution*. London: Weidenfeld &

Nicholson.

Matutinovic, Igor. 2002. "Organizational Patterns of Economies: An Ecological Perspective." *Ecological Economics* 40, no. 3: 421–40.

——. 2006. "Mass Migrations, Income Inequality and Ecosystems Health in the Second Wave of a Globalization." *Ecological Economics* 59, no. 2: 199–205.

Max-Neef, Manfred A. 2005. "Foundations of Transdisciplinarity." *Ecological Economics* 53, no. 1: 5–16.

Naredo, José Manuel. 1987. *La economía en evolución: Historia y perspectivas de las categorías básicas del pensamiento económico*. Madrid: Siglo XXI.

——. 2006. *Raíces económicas del deterioro ecológico y social*. Madrid: Siglo XXI.

Neurath, Otto. 1973. *Empiricism and Sociology*. Edited by Marie Neurath and Robert S. Cohen. Translated by Paul Foulkes and Marie Neurath. Dordrecht: Reidel.

Nielsen, Søren Nors. 2007. "What has Modern Ecosystem Theory to Offer to Cleaner Production, Industrial Ecology and Society? The Views of an Ecologist." *Journal of Cleaner Production* 15, no. 17: 1639–53.

Odum, Eugene P., and Fausto O. Sarmiento. 1998. *Ecología: El puente entre ciencia y sostenibilidad.* México D.F.: McGraw-Hill Interamericana. English version: Eugene P. Odum. *Ecology: A Bridge Between Science and Society.* Sunderland, MA: Sinauer Associates, 1997.

Ostrom, Elinor. 1990. *Governing the Commons: The Evolution of Institutions for Collective Action.* The Political Economics of Institutions and Decisions Series. Cambridge: Cambridge University Press.

——. 2009. "A General Framework for Analyzing Sustainability of Social-Ecological Systems." *Science* 24, no. 325 (July 24): 419–22.

Paavola, Jouni. 2007. "Institutions and Environmental Governance: A Reconceptualization." *Ecological Economics* 63, no. 1: 93–103.

Polimeni, John M., Kozo Mayumi, Mario Giampietro, and Blake Alcott. 2008. *The Jevons Paradox and the Myth of Resource Efficiency Improvements.* London: Earthscan.

Rammel, Christian, and Markus Staudinger. 2002. "Evolution, Variability and Sustainable Development." *International Journal of Sustainable Development & World Ecology* 9, no. 4: 301–13.

——. 2004. "The Bridge Between Diversity and Adaptivity: Answering McIntosh and Jeffrey." *International Journal of Sustainable Development & World Ecology* 11, no. 1: 9–23.

Rammel, Christian, and Jeroen C.J.M. van den Bergh. 2003. "Evolutionary Policies for Sustainable Development: Adaptive Flexibility and Risk Minimising." *Ecological Economics* 47, nos. 2–3: 121–33.

Rammel, Christian, Sigrid Stagl, and Harald Wilfing. 2007. "Managing Complex Adaptive Systems—A Coevolutionary Perspective on Natural Resource Management." *Ecological Economics* 62, no. 1: 9–21.

Rihani, Samir. 2002. *Complex Systems Theory and Development Practice: Understanding Non-linear Realities.* London: Zed Books.

Ring, Irene. 1997. "Evolutionary Strategies in Environmental Policy." *Ecological Economics* 23, no. 3: 237–49.

Schütz, Joachim. 1999. "Organising Diversity." In *Sustainability in Question: The Search for a Conceptual Framework,* edited by Jorg Kohn, John M. Gowdy, Fritz Hinterberger, and Jan Van Der Straaten. Advances in Ecological Economics Series. London: Edward Elgar.

9

Environmental Values, the Epistemology of Complex Problems, and Postnormal Science in the Face of Global Change

Ignacio Ayestarán
The University of the Basque Country (UPV/EHU)

Techno-Scientific Innovations and Planetary Limits

The world has become interconnected at an accelerated pace in the last few centuries, and still more so in the last few decades. In the early-sixteenth century, Magellan and Elcano needed three years to circumnavigate the world. In the late-twentieth century, a jet could make that same planetary tour by air in only twenty-four hours. In the early-twenty-first century, a spacecraft makes one revolution around the planet approximately every hour and a half.

An entire sequence of technological and scientific innovations has shaped the global interconnection of our world, in six successive waves from the Industrial Revolution of the eighteenth century to the present. The first wave of techno-industrial innovation proceeded from ironworking, hydraulic machinery, the mechanization of the textile industry, and the great development of commerce associated with these inventions between the late-eighteenth and mid-nineteenth centuries. The second wave was driven by the steam engine, the railroad, steelmaking, and the cotton industry in the second half of the nineteenth century. At the beginning of the twentieth century and over the course of half a century came the wave of electricity, chemistry, and the internal-combustion engine. In the mid-twentieth century, petrochemistry, electronics, aviation, and the conquest of space initiated the next wave of innovation. The fifth wave culminated in the late-twentieth century with digital networks, biotechnology, computer science, and information and communications technologies. Now in the early-twenty-first century, the sixth wave is making its appearance as the option for a new world based on sustainable production and technologies, system design, biomimesis, green chemistry, industrial ecology, renewable energy, and green nanotechnology, among other innovations visible on this century's horizon.

These technological and industrial changes and innovations have transformed the indexes and factors relevant to the earth's ecosystems on a global scale. Johan Rockström and a broad interdisciplinary team of scientists have detected Earth's limits and critical thresholds that humanity should respect in order not to destabilize essential terrestrial systems. Such disruptions could generate abrupt nonlinear changes. They have identified the following new key processes in the planet's dynamics (Rockström et al. 2009, 473): climate change, rate of biodiversity loss, interference with the global nitrogen and phosphorous cycles, stratospheric ozone depletion, ocean acidification, global freshwater use, changes in land use, chemical pollution, and atmospheric aerosol loading.

Three of these processes have already transgressed their boundaries: climate change, rate of biodiversity loss, and interference with the nitrogen cycle. Four others are also close to doing so: global freshwater use, change in land use, ocean acidification, and interference with the phosphorous cycle. The resulting changes have had such an impact on the planet's limits that some scientists go so far as to affirm that we have altered the geological chronology of the Quaternary Period and that the Holocene has been transformed into a new epoch, the *Anthropocene*, in which humanity has emerged as a global geological force capable of modifying Earth's surface and atmosphere (Steffen, Crutzen, and McNeill 2007; Ayestaran 2008).

Complex Environmental Problems and Postnormal Science: Soft Facts and Hard Values

The assumptions that go into these limits show the difficulties of research to develop robust methodologies in response to global change. While science was previously represented as a continuous progress toward certainty and control over our natural world, it now has to deal with many uncertainties in public-policy issues related to risks and the environment. The analytical, reductionist, and mechanistic paradigm that divides social and ecological systems into ever-smaller parts is being replaced by a systemic approach that integrates reflexive complexity. It is now recognized that natural systems are complex and dynamic, and likewise that those associated with human beings are emergent.

The knowledge appropriate to these new conditions will be based on assumptions of unpredictability, incomplete control, and a plurality of legitimate perspectives extending beyond the laboratory science of Thomas S. Kuhn, who defined the paradigm of "normal science" in terms of three conditions situated in the realm between empiricism and theory: determining the significant facts, matching the facts to the theory, and articulating the theory (1970, 24). Nevertheless, experimentation today is no longer limited to the laboratory because the impacts of complex and global environmental problems affect citizens in multiple ways. For this reason, it is useful to turn to the scenario of postnormal science (Ayestaran and Funtowicz 2010).

In order to characterize the model of postnormal science as applied to problem-solving strategies in response to complex environmental risks, it must be kept in mind that the facts are uncertain, the values in dispute are high, and the decisions are urgent. As a result, a simple linear methodology based on "pure" laboratory science is probably not going to provide a secure guide. This does not imply denigrating the contributions of traditional science, but it does entail a new scientific and methodological strategy that can be graphically represented along two axes (Funtowicz and Ravetz 1993) (figure 9.1). The diagram of postnormal science reveals three distinctive characteristics depicted in terms of two axes. The first presupposes an innovation in scientific methodology and shows the interaction of the epistemic (knowledge) and axiological (values) aspects of scientific problems. These aspects are positioned as axes of the diagram, respectively representing the degree of uncertainty and the scale of the decision stakes in play. The second innovative characteristic is that the uncertainty and the decision stakes in play are the opposites of the attributes that have traditionally characterized science, namely its certainty and value neutrality. Finally, the third distinctive characteristic is reflected in the fact that each of these dimensions is depicted as encompassing three discrete intervals (applied science, professional consultancy, and postnormal science), with three zones that represent and characterize the same number of types of problem-solving strategies.

Figure 9.1. Diagram of post-normal science.

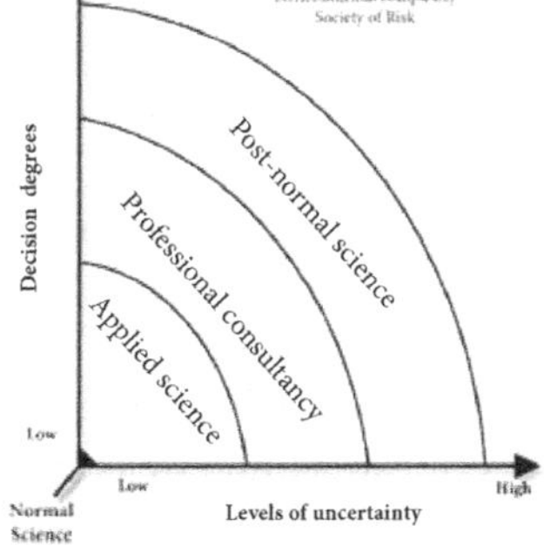

Figure 9.1's two axes are the systems' uncertainties and the decision stakes in play. The axis representing the level of systems uncertainties com-

municates the principle that the problem is not the discovery of a particular fact, but rather the comprehension or management of a reality about which our knowledge is inherently complex. The axis of the decision stakes in play includes all of the various costs, benefits, and value commitments that the problem involves through the various people who intervene and are at risk in the decisions. A small, darkened area has been left at the intersection of the two axes to represent that place that would be occupied by Kuhn's normal laboratory science.

Science has changed its approach to uncertainty in contexts of complexity that outstrip the test tubes of the traditional experimental lab. The conditions of absolute stability and control fundamental to basic scientific research are partially exceeded in applied science, although the level of uncertainty can still be managed at a technical level through standard routines and procedures, statistical tools, and data-processing packages. When uncertainty cannot be managed in this way, personal judgments dependent on high-level skills are required, and uncertainty appears on a methodological level in the midst of increasingly high stakes decision making. At this point, it becomes necessary to appeal to professional consultancy, because greater creativity as well as a readiness to grasp new and unexpected situations and take responsibility for the results are demanded, delimiting the sphere of the design and engineering professions.

Professional consultancy represents an additional step in knowledge. Pure and applied science are characterized by reproducibility and predictability, because they operate in isolated and controlled natural systems. Professional activities, by contrast, deal with unique situations. Professional consultancy is part of the dynamic of expertise and counter-expertise, with a single opinion being insufficient. In these circumstances, the judgments of experts are as important as experimental or field data. In this way, if in pure and applied science the focus of evaluation is the product, it now extends to the product, the process, and the people (triple "p" cubed), because objective and simple criteria or processes for quality assurance are impossible.

Finally, in the most open part of the diagram we find the interval of postnormal science, where professional tasks or applied research exercises can no longer dominate the decision-making process. In this phase, complex uncertainties, new criteria for quality, and sociopolitical commitments arise. The methodological, social, and ethical considerations exceed the traditional framework of the scientific enterprise. The normal processes for project review by evaluators and the academic weighing of articles are no longer sufficient, because the community of researchers has expanded to include administrators, users, affected individuals, and citizens in general. The extension of the community of evaluators may come to encompass scientists from a variety of disciplines, including lawyers, journalists, consumers, and pressure groups that intervene in politics marked by complex contexts and high levels of uncertainty, because the decision stakes are prolonged in space and time. The problems of environmental and social risk introduce new demands that a simplistic and reductionist vision of scientific practice cannot address.

In postnormal science, the search for quality has the consequence that axiological and epistemological neutrality cannot be presupposed. On the contrary, it is necessary to express and make explicit the values and uncertainties that play a role in decisions in order for those decisions to be effective and reasonable. In dialogue about a problem of postnormal science, the primary guide is the quality of the stakeholders and not the truth of only one of the parties involved. Likewise, the unity of postnormal science does not derive so much from a shared knowledge base as from a shared commitment by the parties and their various methodologies and techniques of knowledge. Research tools in the postnormal phase include processes for discussion (e.g., focus groups and expert panels) and for deliberation (e.g., mediation, juries, or citizen parliaments and consensus meetings), without excluding debates and publicity in the communications media and making use of information technology. Postnormal evaluation thus achieves various objectives from the starting point of the different stakeholders involved, including quality of the results, legitimacy of the processes, approval by public opinion, and the spread of good citizenship and democracy.

This postnormal science has the paradoxical characteristic that in its problem-solving activity it inverts the traditional dominance of "hard facts" over "soft values," as well as the division between "hard sciences" and "soft sciences," and technical versus nonexpert knowledge. Traditional scientific inputs have become "soft" in the complex context of "hard" value commitments. The traditional and neo-positivist division between "facts" and "values" is inverted because the two categories cannot realistically be separated. The uncertainties go beyond the systems to include ethics and politics. In global environmental risks, there are no longer factors external to the techno-scientific enterprise, but rather an expansion of the evaluative community in response to potential risks. When problems do not have clear solutions, when their environmental and ethical aspects are fundamental, when the phenomena themselves are ambiguous, and when all research techniques are open to methodological criticism, then excluding the entire world other than official experts and specialized researchers with academic or administrative posts does not intensify debates about quality.

In this way, research activity more closely resembles the functioning of a democratic society, characterized by expanded participation and a marked tolerance for diversity. From the perspective of postnormal science, recognition of global environmental risks demonstrates that the appropriate framework for problem solving is not that of universal rationality, but rather of "glocal" (simultaneously global and local) reasonableness, bringing together diverse methodologies (axiology, epistemology, and politics) that entail different areas and registers (table 9.1)

Table 9.1. Postnormal science: from universal rationality to glocal reasonableness.

Methodologies	Axiology	Epistemology	Politics
Areas involved	Decision stakes	Systems uncertainties	Vulnerability scenarios
Registers	Values evaluated	Expanded participation	Associated risks

This new paradigm of postnormal science presents itself as a possible governance model for risk at the convergence of three axes: axiology (decision stakes), politics (vulnerability scenarios), and epistemology (systems uncertainties). Figure 9.2 charts the evolution of synergies in knowledge through a new social contract for science and technology (Ayestaran and Funtowicz 2010).

Figure 9.2. Evolution of synergies in knowledge.

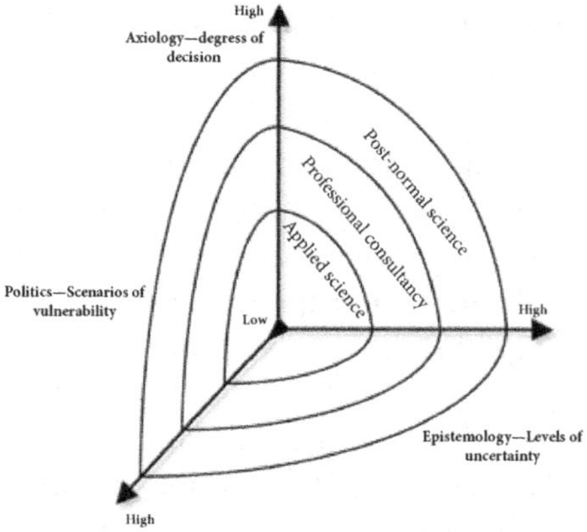

Sustainability Science: The Post-Kuhnian Paradigm of Sustainability

In parallel to the methodologies of postnormal science, there has been an effort by various scientists in recent years to constitute an interdisciplinary field under the name of *sustainability science*, with the aim of overcoming the limited efforts situated between global uncertainty and global knowledge. Along these lines, in 2001 scientists from more than 100 countries adopted the Amsterdam Declaration on Global Change, which formally established the Earth System Science Partnership. This declaration acknowledged a consensus around five points (Moore III et al. 2002):

1. The Earth System behaves as a single, self-regulating system, with physical, chemical, biological, and human components.
2. Human activities are significantly influencing Earth's environment in many ways, in addition to greenhouse gas emissions and climate change.
3. Global change cannot be understood in terms of a simple cause–effect paradigm.
4. Earth System dynamics are characterized by critical thresholds and abrupt changes. Human activities could inadvertently trigger such changes with severe consequences for Earth's environment and inhabitants.
5. In terms of some key environmental parameters, the Earth System has moved well outside the range of the natural variability exhibited over the last half million years at least.

On this foundation, the declaration sought unanimity on two demands: (1) an ethical framework for global stewardship and strategies for Earth System management are urgently needed, and (2) a new system of global environmental science is required.

Following this declaration, the aim was to consolidate a program that would evaluate the methodological issues related to the emerging paradigm of Earth System science. Since then, this scientific community has developed its own "Hilbertian program" (Clark, Crutzen, and Schellnhuber 2004, 8–14; Costanza, Graumlich, and Steffen 2007, 420). Nevertheless, the new sustainability science implies a post-Kuhnian paradigm model and one that belongs to postnormal science. Kuhn (1970, 176) specifies that a paradigm is what is shared by the members of a scientific community, and, conversely, a scientific community is made up of people who share a paradigm. Appealing to the communitarian structure of science, he notes two different senses of paradigm: (1) the constellation of beliefs, values, techniques, and other elements shared by the members of a given community, and (2) the concrete solutions to puzzles that, used as models or examples, can substitute for explicit rules as a basis for solving the remaining puzzles of normal science. This methodological strategy leads Kuhn to maintain the thesis that paradigms are disciplinary matrices. The paradigmatic elements of a disciplinary matrix are as follow (Kuhn 1970, 182–87):

1. *Symbolic generalizations:* the formal or formalizable components, symbols, and words that the members of a group deploy without dissent. They may be in the form of natural laws or definitions.
2. *Metaphysical models:* the shared commitments to and beliefs in particular models, analogies, and metaphors that function as heuristic

models. The metaphysical components provide the heuristics and ontology (of the phenomena and objects) of a disciplinary matrix.

3. *Values:* values about predictions, the evaluation of theories, and the social utility of science (Kuhn would subsequently reformulate his list of values in different terms).
4. *Exemplars:* the examples shared by the group of scientists in solving problems, ranging from the examples learned by a student in his or her scientific education to the examples found in specialized periodicals. Exemplars provide "the fine structure of science" and are for Kuhn the predominant element in a disciplinary matrix.

Following this scheme, the transition to the disciplinary matrix of a sustainability–science paradigm could be characterized as follows:

1. *Symbolic generalizations:* the epistemic principles accepted by population ecology, the (neo-)Darwinian evolutionary synthesis, and the laws of thermodynamics, in addition to other generalizations originating in economics and the social sciences. The Hilbertian program of Earth System science would also be an epistemic foundation for this part of the matrix.
2. *Metaphysical models:* the concept of "system" functions as an ontological and heuristic commitment, seeking a functional and dynamic unity among interrelated parts that will structure research about the natural and social components under examination.
3. *Values:* the predominant value is system resilience along the three dimensions of society, the economy, and the environment in the interaction between natural and human systems. Other ethical and social values also exist, such as those advocated in the Amsterdam Declaration.
4. *Exemplars:* the study of abrupt anthropogenic climate change functions as an overarching exemplar of sustainability science, although it appears together with other exemplars (exploitation of hydrologic systems and desertification, loss of biodiversity, land management, and urban and demographic planning, among others).

The specific characteristics of sustainability science fit globally into Kuhn's proposed framework for paradigm matrices; nevertheless, divergences also exist on the following points.

Kuhn's model of a disciplinary matrix refers to an established model of "normal science," while sustainability science is more appropriately located in the realm of "trans-science" (Weinberg 1972), the space where science poses questions that science alone cannot answer, as in the model of "postnormal science" proposed by Funtowicz and Ravetz.

The matrix of normal science in Kuhn's model seeks the solution to problems conceived as puzzles (Kuhn 1970, 35–42), while sustainability science aims to manage problems and risks, taking as its guiding principle quality of life and not only problem-solving efficiency, in the form of a new heuristic.

Kuhn conceives of normal science as a model for a *restricted scientific community*, while in postnormal sustainability science this model is expanded to an *extended peer community* with an epistemic gradient of different levels of knowledge, uncertainty, and decision stakes.

Revolutions in normal science, in Kuhn's view, are entirely distinct from political ones, while sustainability science is equally a scientific revolution and a *techno-political* one, with a new axiology about the social value of science, progress, and our coevolving environment.

Kuhnian normal science has a disciplinary matrix, while the expanded knowledge of sustainability science is transdisciplinary. The structure of scientific revolutions was initially conceived for the disciplinary science of the classical natural sciences, while the silent revolution of sustainability science presupposes a hybrid transdisciplinarity between the natural and social sciences.

In the Kuhnian model of normal science, the publication and dissemination of scientific studies takes places through classical basic sources: scientific textbooks, popularizing works, and the philosophy of science. In the post-Kuhnian model of sustainability science, the communication and publication of scientific studies expands the methods of normal science, employing new hybrid forums of participation, new mass-media possibilities, and new information and communications technologies.

In summary, the rise of sustainability science constitutes a new paradigm. It is born as a hybrid of other older paradigms, as well as one that still needs to seek a language of its own in relation to the concept of "sustainability" (Norton 2005). This clearly exceeds the frameworks of normal science in the direction of other postnormal models in epistemology, axiology, and political activity.

The Complexity of Inverse Problems and Wicked Problems

The motive for postnormal science to bring to light the need to expand knowledge in the context of uncertainty that surrounds complex environmental problems and to demand a paradigm shift in the approach to sustainability lies in its formulation, from an epistemological perspective, of inverse problems. According to Mario Bunge (2003, 203–6), in managing and investigating problems, we face two classes of problems, direct and inverse, and it is useful to distinguish the one from the other. Direct problems run downstream, from premises to conclusions or from causes to effects. Inverse problems, on the other hand, entail an inversion of the usual logical or causal chain. For this reason, a peculiarity of inverse problems is that they may have multiple solutions or several plausible scenarios for their resolution.

In resolving a direct problem, we seek to discover its behavior, given a system's composition, context, and structure. To resolve the corresponding inverse problem, we rely on inference to surmise its composition, context, and structure. In addition, inverse problems are much more difficult to resolve, because it may be the case that different internal mechanisms produce the same behavior or the same apparent effect.

The resolution of a direct problem involves analysis or progressive reasoning, whether from premises to conclusions or from causes to effects. The resolution of an inverse problem, by contrast, involves synthesis or regressive reasoning, either from conclusions to premises or from effects to causes. In general, direct problems can be resolved, even if only approximately. Inverse problems, on the other hand, rarely have a single, clear, and definitive solution.

Generally, when a direct problem is posed, the raw material or cause (input) is given, and the product or effect (output) is the subject of investigation. In the inverse case, the product or effect is given, and the raw material or cause is what is being sought. In both cases, hypotheses that relate causes to effects are used or developed, but the direction of research alters the degree of solution, especially in cases of diagnosis. From this perspective, direct and inverse problems might be represented as follows:

1. *Direct problem:* given $C \rightarrow E$ and C, find E, where $\rightarrow$ symbolizes the causal relationship. For example, find the effect on behavior of the ingestion of a pharmaceutical (or a chemical product) administered (or emitted) in a specific dose.
2. *Inverse problem:* given a clinical (or ecological–environmental) datum E and the plausible causal hypotheses $C_1 \rightarrow E$, $C_2 \rightarrow E$, ... , $C_n \rightarrow E$, find cause number 'i' that has played a role.

Many inverse problems are diagnostic problems that arise in the search for solutions. The problems of global sustainability commonly appear in the form of inverse problems in at least two respects: in their detection and in their solution. First, their detection leaves room for several plausible scenarios as causes or inputs. In this way, for example, anthropogenic climate change will always be a topic for debate, despite the detected increase in temperatures, since we have to work backward to possible previous scenarios, about which complete information is by definition lacking. Second, an inverse problem almost always appears in the search for solutions to problems of nonsustainability, since they are going to correspond to different management frameworks with different possible solutions, which will generate new problems in their turn in the future. There will never be a single definitive solution that can be subjected to an algorithm or mechanical rule, just as what happened with the paradigm of postnormal science.

In the early 1970s, it was realized that in urban and social planning there are two classes of problems (Rittel and Webber 1973): one group of "tame" or "benign problems" and another group of "wicked problems." This second group is made up of problems characterized as "wicked" (in contrast to "benign"), "vicious" (as in a "vicious circle"), "tricky" (like a trickster spirit), and "aggressive" (like a lion, in contrast to a lamb's docility). Benign and tameable problems occur more frequently in the natural and exact sciences. Examples are a mathematician faced with a problem that can be solved with an equation, or a chess player who can force checkmate in five moves, where the problems can be defined and isolated and can have localizable solutions. Tangled and twisted problems, in contrast, are difficult to define, isolate, or delimit and are ubiquitous or of difficult localization, as is the case in the majority of problems in public or social planning (e.g., the problems faced by planners and managers in building a new highway). Benign and tameable problems are solved, while twisted and perverse problems are resolved but not dissolved (for which reason they have to be *re*-solved again and again).

Expanding the planning perspective on wicked problems (Rittel 1972, Rittel and Weber 1973), we can identify 10 fundamental characteristics of this class of complex problems:

- Unlike a tameable problem, a twisted or perverse problem has no definitive formulation; even the definition and scope of the problem are controversial, because each question depends on the state of its solution at a given time, and the next step cannot be foreseen.
- In contrast to tameable problems, wicked problems lack a definitive solution. In a tameable problem in chess or mathematics, the stopping point is known. For example, checkmate can be achieved in three moves, and the correct combination is then known. Or we can also know that if something like "x = y" is reached in a particular equation, then we know that we have gotten it right. In a perverse problem, we generally stop because we are out of time, money, patience, or all three.
- The solutions to wicked problems are not true or false, with mistakes and errors marked on a truth table. An urban plan is not right or wrong, but good or bad, appropriate or inappropriate, to different degrees and in the eyes of interested parties. As a consequence, a perverse or twisted problem is not susceptible of solution by the usual epistemic criterion of true/false, but rather presupposes a

plurality of approaches, capabilities, and values. The more complete this approach is for the involved and interested parties, the better the solutions reached will be, although they will never be definitive.

- Not coincidentally, there is no immediate or final or definitive proof of a solution to a wicked problem, since each action taken to counteract it will have consequences over time (next month, next year, or next decade). There may always be additional potential consequences, and for that reason, there is no final corroboration that declares the problem closed.
- In a tameable problem, there is room for several tries: a chess problem can be played again and again, and a mathematical equation can be attempted over and over. Once a tameable problem has been solved, it has been solved generally and will be a prototypical solution. Nevertheless, every attempt to solve a twisted or perverse problem is unique, since the results cannot easily be undone, nor can they be repeated at another time and place, so that no clear opportunity exists to learn by trial and error, unless new problems and scenarios are generated.
- Wicked problems do not have a clear set of potential solutions, nor does there exist a well-described series of permissible operations that could be incorporated into the plan. For this reason, they are not reducible to a mechanical rule or to an algorithm, something that always introduces uncertainty about whether action is being taken on the right scale and in the direction appropriate to the problem. Ethical values and imagination will necessarily enter into play, because no exhaustive list enumerating permitted operations exists.
- In contrast to tameable problems, which are frequently general and routine, every twisted or perverse problem is essentially unique, with the consequence that past experience can only be taken advantage of in the future in a limited way. The old solution may not be effective in the future, depending on opportunity and context.
- Every wicked problem can be considered as a symptom of another problem, because it will frequently be associated with new problems. For that reason, there is always room for doubt about the problem's degree of resolution and scale of management. In this way, yesterday's solutions are today's problems, and today's solutions will be tomorrow's problems. Solutions in one place may also generate problems someplace else.
- The existence of a discrepancy representing a twisted or perverse problem may have numerous explanations, and there is often no evidence that can allow us to discover which explanation is best. Often,

the direction taken by the solution depends in the first instance on the explanation, and consequently, on how the diagnosis of the problem was defined and initiated ("why does the problem exist, and which is it?").

- Unlike those who solve tameable problems, where the actors are not directly culpable for their inability—someone can lose a chess game without being blamed for it, or maintain a hypothesis that is refuted by others without injury other than to personal pride—the planner managing perverse or twisted problems has no right to make a mistake, because that person is responsible for what he or she is doing and, as a consequence, there is no public tolerance for planning experiments that fail. In a democratic context, social and public planning is not subject to trial and error.

This kind of complexity exists in the field of sustainability. Kuhn (1970) says that in the paradigm of normal academic science, problems are solved in the way that someone might solve a puzzle or play chess. Now, in problems such as these examples (puzzles or chess), we have direct problems, not complex or twisted or perverse ones, with the possibility of introducing algorithms and stopping rules. Nevertheless, postnormal science has already pointed out that complex environmental problems entail an expanded model of science and a greater cognitive democratization in both the detection of problems and their solution, since there will generally not be a single perspective. Likewise, the problems associated with a lack of sustainability cannot be remedied by simple solutions or recipes, since they present inverse diagnostic problems, as well as wicked planning and management problems. This entails different degrees of public deliberation and social management in contexts of democratic participation, or at least of open and plural participation, both on the part of scientists and political actors and on that of affected individuals and the body of citizens.

With regard to the wicked problems associated with sustainability, Bryan G. Norton maintains that big environmental and social problems have four inevitable characteristics (2005, 133–36): "problems with problem formulation" are frequent; noncomputable solutions are common; environmental problems are nonrepeatable; and the effects are open, variable, and intertemporal.

The big problems of sustainability are triply perverse and twisted: first, in their definition (always problematic); second, in their diagnosis (where there are inverse problems with plausible scenarios); and, third, in their management and solution (with multiple possibilities and undesired or unforeseen effects over the short or long term). For this reason, they cannot be solved unilaterally or in a limited way. The only path is deliberation, research, shared management, and the democratization of problems and their solutions. For this purpose, it is an indispensable requirement to adopt a rational and methodological pluralism, a greater integration of the interrelationships among science, technology, and society, especially the full range of civil society, and, finally, an ethics of the complexity associated with inevitable uncertainty that can make us more effective and less arrogant.

Environmental Values in a Coevolutionary and Dynamic Context

In line with what has been discussed thus far, a comprehensive and dynamic vision becomes necessary, bringing together knowledge, values, the environment, organizations, and technology. This vision has to be fundamentally *coevolutionary* (Norgaard 1994; Kallis and Norgaard 2010), because it is the one that can best respond to the interconnected dynamics of deliberate innovation, chance discovery, and random change taking place among and within social and ecological systems, as depicted in figure 9.3.

Figure 9.3. Interconnected dynamics.

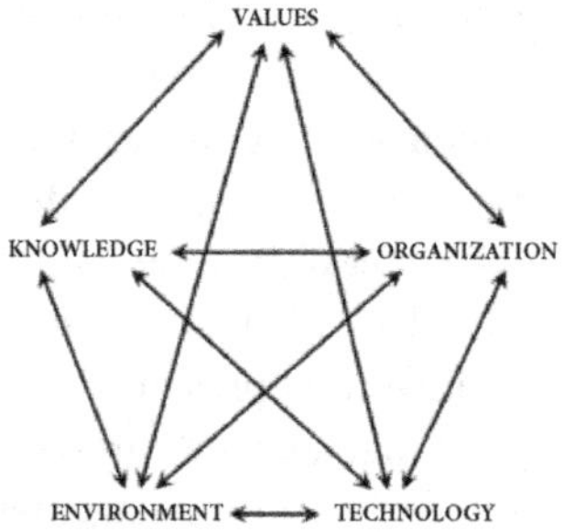

For this purpose, it is necessary to conciliate environmental values with evaluative pluralism, leading us to some reflections about environmental values and their construction and public explication. It is useful to recall a pragmatic analysis of environmental values (O'Neill, Holland, and Light 2008, 1):

- There is no such *thing* as the environment. *The* environment—singular—does not exist. In its basic sense, to talk of the environment is to talk of the environs or surroundings *of* some person, being, or community. In practice, to talk of the environment is at best a shorthand way of referring to a variety of places, processes, and objects that matter, for good or bad, to particular beings and communities: forests, cities, seas, weather, houses, marshlands, beaches, mountains, quarries, gardens, roads and rubbish heaps.
- There are no such *things* as values. Rather, there are the various ways in which individuals, processes, and places matter, our various

modes of relating to them, and the various considerations that enter into our deliberations about action.

- Environments—plural—and their constituents, good and bad, matter to us in different ways.
  - We live *from* them: they are the means to our existence.
  - We live *in* them: they are our homes and familiar places in which everyday life takes place and draws its meaning, and in which personal and social histories are embodied.
  - We live *with* them: our lives take place against the backdrop of a natural world that existed before us and will continue to exist beyond the life of the last human, a world that we enter and for which awe and wonder are appropriate responses.

Values are vectors of attention, decision, and realization that guide our knowledge, habits, and behaviors in response to both social and natural environments, in varied matrices. We live from the world, we live in the world, and we live with the world in coevolution. For this reason, there is not an isolated nature from which environmental values emanate; rather, environmental values are found in these relationships with, in, and from our interaction with social systems and ecosystems. This has the consequence that values are more plural, but also that they are not exempt from conflict or debate. Axiological pluralism, open deliberation as part of public decisions, and the new methodologies of complex problems are indispensable for this interdisciplinary debate on global problems such as climate change. As a scientist on the Intergovernmental Panel on Climate Change used to tell the story, when that panel began to meet, around twenty years ago now, an elderly Japanese scientist in the group spoke up during one of the meetings and said: "We scientists have confirmed that there is an emissions problem, but we can't solve it. Since CO_2 is produced by machines, we will have to call in the engineers. They, in turn, will say that the technology necessary to solve the problem exists, but that it costs money, and so the economists will be called in. The economists will do their calculations and will say that, in order to achieve this, it will be necessary to change our current social model based on transportation, on squandering energy—and so the sociologists will be called in. They, in turn, will say that it's a value-scale problem that they can't solve, and so the philosophers will be charged with telling us what values we should emphasize and be interested in."

References

Ayestaran, Ignacio 2008. "The Second Copernican Revolution in the Anthropocene: An Overview." *Revista Internacional de Sostenibilidad, Tecnología y Humanismo* 3: 145–57.

Ayestaran, Ignacio, and Silvio O. Funtowicz. 2010. "Ciencia postnormal, problemas ambientales complejos y modelos de información." *Ludus Vitalis* 16, n. 33: 25–48.

Bunge, Mario. 2003. *Emergence and Convergence: Qualitative Novelty and the Unity of Knowledge*. Toronto: University of Toronto Press.

——. 2006. *Chasing Reality: Strife over Realism*. Toronto: University of Toronto Press.

Clark, William C., Paul J. Crutzen, and Hans Joachim and Schellnhuber, eds. 2004. "Science for Global Sustainability: Toward a New Paradigm." In *Earth System Analysis for Sustainability*, edited by Hans Joachim Schellnhuber, Paul J. Crutzen, William C. Clark, Martin Claussen, and Hermann Held. Cambridge, MA: The MIT Press and Dahlem University Press.

Costanza, Robert, Lisa J. Graumlich, and Will Steffen, eds. 2007. *Sustainability or Collapse? An Integrated History and Future of People on Earth*. Cambridge, MA: The MIT Press and Dahlem University Press.

Funtowicz, Silvio O., and Jerome R. Ravetz. 1993. "Science for the Post-normal Age." *Futures* 25, no. 7: 739–55.

Kallis, Giorgos, and Richard B. Norgaard. 2010. "Coevolutionary Ecological Economics." *Ecological Economics* 69, no. 4: 690–99.

Kuhn, Thomas S. 1970. *The Structure of Scientific Revolutions*. Chicago: University of Chicago Press.

Moore III, Berrien, Arild Underdal, Peter Lemke, and Michel Loreau. 2002. "The Amsterdam Declaration on Global Change." In *Challenges of a Changing Earth*, edited by Will Steffen, Jill Jäger, David J. Carson, and Clare Bradshaw. Berlin: Springer-Verlag.

Norgaard, Richard B. 1994. *Development Betrayed: The End of Progress and a Coevolutionary Revisioning of the Future*. London: Routledge.

Norton, Bryan G. 2005. *Sustainability: A Philosophy of Adaptive Ecosystem Management*. Chicago: The University of Chicago Press.

O'Neill, John, Alan Holland, and Andrew Light. 2008. *Environmental Values*. London and New York: Routledge.

Rittel, Horst W.J. 1972. "On the Planning Crisis: Systems Analysis of the 'First and Second Generations.'" *Bedrifts Økonomen* 8: 390–96.

Rittel, Horst W.J., and Melvin M. Webber. 1973. "Dilemmas in a General Theory of Planning." *Policy Sciences* 4: 155–69.

Rockström, Johan et al. 2009: "A Safe Operating Space for Humanity." *Nature* 461 (September 24): 472–75.

Steffen, Will, Paul J. Crutzen, and John R. McNeill. 2007: "The Anthropocene: Are Humans Now Overwhelming the Great Forces of Nature?" *Ambio* 36, no. 8: 614–21.

Weinberg, Alvin 1972: "Science and Trans-Science." *Minerva* 10, no. 2: 209–22.

10

Science, Gender, and Sustainable Development

Teresa Nuño Angós

The University of the Basque Country (UPV/EHU)

The Genesis of the Concept of Sustainable Development

Since the United Nations (UN) Conference on the Human Environment in Stockholm in 1972, the major conferences of the UN have been full of declarations and intentions concerning the environment. Today, everyone (including the major energy companies) talks about "sustainable development," an expression that has turned into an umbrella term that encompasses very different conceptions and that ought to be clarified in order to properly advance any critical form of scientific and environmental education. The United Nations World Commission on Environment and Development (1983–87), chaired by Gro Harlem Bruntland, defines *sustainable development* as follows: "A process of change in which the exploitation of resources, the direction of investments, the orientation of technological development and institutional change are all in harmony and enhance both current and future potential to meet human needs and aspirations" (WCED 1987, 57).

A Review of the Concept of Sustainable Development Using Gender as a Category of Analysis: A Holistic Concept

From a gender perspective, the First International Conference on Women was held in 1975 in order to promote greater equality between men and women and to put an end to the marginalization and oppression of the female majority. Even before that year, the pioneering study carried out by Ester Boserup (1970) challenged the development policies in vogue at the time by demonstrating the importance of the economic participation of women and their contribution to the development of "backward" societies. Her research not only questioned the assumptions of traditional conceptions of development (which ignored women's participation), but also questioned the impact of the development policies implemented, given their failure to take these contributions into account (Alvarez-Lires 2008). In the mid-1950s, the analytical concept of gender became the most fundamental concept underpinning new approaches to women and development. This perspective of *gender* and *development* aimed to put an end to the limitations of previous approaches and to valorize the role of women in development.

In spite of this, gender as a category of analysis is absent to a large extent from the studies carried out in favor of sustainable development, even though it has started to become more apparent in international organizations, thanks to the tenacity of groups of women from the so-called Third World. For example, the awarding of the Nobel Peace Prize to Wangari Maathai of Kenya in 2004, in recognition of her struggle in support of the environment and women, was decisive.

The question of gender represents a challenge on a global scale, but for development NGOs, *comprehensive development* and *sustainable development* are empty words in the context of inequality between human beings. The study of this inequality has led to different perspectives of women within the framework of development theories:

- In the 1950s and 1960s, women were still viewed as passive objects.
- In the 1970s, women began to play a productive role.
- By the mid-1970s, women were considered pillars of economic development.
- In the 1980s, under a more holistic approach, the study of gender relations begins—a study that takes into account not only women, but also men—and women are viewed as subjects of change and struggle as well as active agents of development.

Women in the Environmental Sciences

Women have been systematically discriminated against in the field of science. Nevertheless, some areas of biological sciences have been less reluctant to accept women and some of them have been able to make important scientific contributions in biology, specifically in the field of environmental sciences. Scientists from across the globe are noted for their work in this field, including Lynn Margulis, Ellen Swallow Richards, Rachel Carson, Gro Harlem Brundtland, Wangari Mathaai, Vandana Shiva, Bina Agarwal, and Val Plumwood. The contributions made by these scientists in the field of the environmental sciences are presented and analyzed in the sections that follow.

The Origin of Life: Symbiogenesis and the Gaia Hypothesis

Lynn Alexander Margulis (1938–2011), Chicago, Illinois (USA)

Around 1967, building on the studies of Kostantin Merezhovsky, which had been rejected by official science of the day, Lynn Margulis formulated her theory of serial endosymbiosis (or symbiogenesis). Initially criticized, it is now accepted as a plausible theory that could offer a convincing explanation for the evolution of eukaryotic cells. Margulis (1981) developed her theory on the role played by symbiosis in the origin of the eukaryotic cell. She attempted to explain how two living beings of two different species join together in order to live together, thus resulting in a new species. This theory is based on two biological concepts:

1. The most fundamental division of the organization of the living world is that which exists between prokaryotes (organisms whose cells do not have a nucleus) and eukaryotes (whose cells have a nucleus)—that is, between bacteria (without nucleus) and other organisms made up of nucleated cells (protoctista, plants, fungi, and animals).
2. Some parts of eukaryotic cells emerge directly from the formation of permanent associations between different organisms by symbiosis.

Unlike bacteria, the cells of plants and animals have a nucleus. They also contain organelles, such as mitochondria and plastids whose distant ancestors were bacteria. Thanks to the association of certain bacteria with other bacteria with which they initially established an intracellular symbiotic relation, the first protoctista (eukaryotic organisms) were obtained, and from there the other three kingdoms of eukaryotes emerged: fungi, plants, and animals. The associations occurred over a long period of time and did not occur simultaneously, hence the term *serial endosymbiosis*.

Margulis was convinced that symbiogenesis is a process that occurs much more frequently than is thought by those who support evolutionary theories that continue in Darwin's footsteps, a tradition that emphasizes competition much more than cooperation in the evolutionary process. According to Margulis, these kinds of symbiotic "marriages" are the essence of the evolution of species on all levels. However, she stressed that cooperation is not the same as symbiogenesis. Nobody wins or loses with symbiosis; rather, there is a recombination and something new is made.

The significant contribution made by this scientist affected evolutionist discourse. Margulis demonstrated that the nucleated or eukaryotic cells not only descended from bacteria, but also were literally amalgams of different bacterial cells. As she explained (Margulis 1998), there are four parts to the symbiogenesis of a eukaryotic cell: the origin of the nucleocytoplasm; the cilia (that are also centrioles); the mitochondrial part; and the plastids. The role of each of these has since been proven and documented, except for the part concerning the cilia, which has not been proven entirely and is an aspect that Margulis continued to work on in order to ensure it was duly researched.

While not breaking entirely with the postulates of Darwin and his successors, Margulis managed to demonstrate the value of symbiosis as an evolutionary mechanism. The speciation process is directly linked to the acquisition of symbionts, and this theory, alongside original Darwinism, enables evolution to be explained much more thoroughly. Margulis (1986, with her son, Dorion Sagan) developed her theory of symbiogenesis further and linked it to the theory of evolution.

During the process of developing and verifying her theory, around 1972 she met the British chemist James E. Lovelock, a specialist in atmospheric sciences. Three years earlier, Lovelock had put forward his hypothesis of the earth as an auto-regulating system, more widely known as the Gaia hypothesis. However, what Lovelock did not know was by which mechanism the earth was able to control its temperature and the composition of its atmosphere. He knew that the auto-regulatory processes had to involve live organisms, but he did not know which. Margulis was studying the same processes as Lovelock, researching the production and elimination of gases by different organisms, including bacteria. The collaboration of both scientists was ideal for this research. Together they unveiled a complex network of feedback loops that were responsible for the auto-regulation of the planet. These loops link live systems with nonlive ones. They demonstrated that a close relationship exists between the living parts of the planet (plants, microorganisms, and animals) and the nonliving ones (rocks, oceans, and the atmosphere).

The Gaia hypothesis considers the life of a systemic mode. Initially, the scientific community rejected the hypothesis as being theological. However, Margulis and Lovelock never claimed that planetary auto-regulation was driven by a divine purpose. Journals such as *Nature* and *Science* rejected their publication. Finally, Carl Sagan (her first husband) invited them to publish in his journal, *Icarius,* in 1974 (Alvarez-Lires, Nuño, and Solsona 2003).

With the Gaia hypothesis, Margulis's gaze lifted from the microscopic level to a global one: the earth is an authentic living system, a breathing amalgam on a global scale of organisms and the inanimate physical world. Gaia is the great ecosystem of which the planet's biosphere is made up, the combination of all of Earth's ecosystems. In short, according to Margulis life is not "a thing," but rather a continuous process.

In *Five Kingdoms* (1982, with Karlene Schwartz), Margulis describes and illustrates all of the main groups of organisms that belong to the five kingdoms (monera, protoctista, fungi, plants, and animals), as proposed by Robert H. Whittaker in 1959. The author of more than 100 articles in international journals, Margulis also published around 60 books on science and research, both for secondary education as well as for university study, that deal with the biogeochemical cycles of sulfur. *Acquiring Genomes* (2002) is worthy of particular mention. Many of her ideas, now classics, are captured in secondary education science books, even though Margulis does not appear cited in the majority of these texts. Margulis continued her work researching the world of bacteria in order to rewrite the history of the evolution of the species on Earth through her theory of symbiogenesis, a task that she was unable to complete due to her premature death.

Domestic Economy (Pioneer of Environmental Sciences)

Ellen Swallow Richards (1842–1911), Dunstable, Massachusetts (USA)

Ellen Richards was the first woman to be admitted into the Massachusetts Institute of Technology (MIT), where she studied chemistry, although she was not allowed to obtain a doctorate. She founded the science of domestic economy, a field of study that combined knowledge of cookery, nutrition, water supply, hygiene, and health. Her objective was to understand the environmental dynamics of industrialization and to provide the community (especially women) with the skills necessary to be able to control their own environment. Richards was, in this respect, a pioneer of environmental sciences.

As a result of her efforts, the Women's Education Association of Boston funded a laboratory for women at MIT, where she worked as assistant director alongside professor John Ordway. Beginning in 1876, Richards was a member of the Society to Encourage Studies at Home and was in charge of the science section. She also cofounded in 1881 the American Association of University Women (AAUW), where she worked to promote the training and empowerment of women. From her post as assistant professor, Richards introduced the teaching of biology at MIT and contributed to the foundation of the Woods Hole Oceanographic Institution. She carried out research on water pollution and designed safe systems for the supply of water. Richards was part of the committee of the Naples Table Association for Promoting Laboratory Research by Women, an association that awarded an annual prize for women that consisted of funding a year's research in biology laboratories, such as the Zoological Station of Naples or Woods Hole. After her death, the award became known as the Ellen Richards Prize.

From 1890, Richards focused her career on improving nutrition in the working classes and was involved in the creation of economic restaurants, which provided nutritional foods and instruction on how to prepare them, as well as guidance to hospitals and schools on healthy diets. She advocated the introduction of domestic economy courses in public centers and, in 1899, designed the teaching and certification requirements for this new science. This gave rise to the formation of the American Home Economics Association.

Richards became the first President of the Association and founded the *Journal of Home Economics.* From 1910, she was part of the board of the National Education Association and was responsible for supervising the teaching of domestic economy in public centers. Richards received an honorary doctorate of science from Smith College in 1910.

Although she has many innovative publications to her name (1886, 1896, 1900, 1911), Richards's original and wide-ranging initiatives were not appreciated by the scientific community of the time, which was rapidly separating into different disciplines. Even less appreciated was her insistence on working with women.

Silent Spring and the US Environmental Movement

Rachel Louise Carson (1907–1964), Springdale, Pennsylvania (USA)

Trained in marine biology and zoology, Rachel Carson worked as a zoologist (1932–1935) at the Johns Hopkins University. She gave instruction and carried out research on living marine organisms (1935–1940) through the U.S. Bureau of Fisheries. She was one of the first two women contracted to undertake a scientific post at the bureau.

In 1941, Carson published *Under the Sea Wind*, which was followed in 1951 by *The Sea Around Us*, in which she discusses and denounces the interactions between humans and ocean beings, warning about the growing dangers of large-scale marine pollution. Consequently, she was elected member of the Royal Society of Literature (Great Britain), awarded the National Book Award for works of nonfiction, and granted a Guggenheim fellowship. In *The Edge of the Sea* (1955), Carson focuses on the delicate network of life and shows the interdependence of all living creatures in seashore communities. From 1936 to 1949, she worked as a marine biologist for the U.S. Fish and Wildlife Service and held the post of editor-in-chief of publications for eleven years (1940–1951). From 1952 until her premature death in 1964, Carson was a member of the Marine Biological Laboratory Corporation.

Her most well-known book, *Silent Spring* (1962), provoked considerable controversy and social impact. The book generated global awareness of the dangers of environmental pollution as well as social concerns for the problems caused by modern synthetic pesticides (especially DDT) and their effects on the food chain, particularly those of birds. Prior to the book's publication, few people realized that pesticides permeated the food chain, increasing their concentration by moving from one trophic level to the next. Through her book, Carson faced up to the chemical giants who manufactured pesticides such as DDT as well as some agencies of the US government. Furthermore, she questioned industrial society's right to contaminate at will, without taking into account the effects on the environment (Alvarez-Lires, Nuño, and

Solsona 2003). The multinationals invested fortunes trying to discredit her, branding her as an alarmist, hysteric, fanatic, and even a Communist. In the end, Carson's book led to the regulation of the manufacture and use of chemical pesticides, as well as the elimination of their waste, and the use of DDT was finally prohibited in the 1970s and 1980s in most industrialized countries.

In spite of the whole campaign unleashed against her, Carson received recognition for *Silent Spring* during the same year that she was elected member of the American Academy of Arts and Letters. In her last book, *The Sense of Wonder* (published posthumously in 1965), she encourages adults to stimulate children's innate capacity to admire nature and marvel at it. In short, her critique of scientific approaches to the natural world presaged the subsequent critiques advanced by ecofeminists: "As man proceeds toward his announced goal of the conquest of nature, he has written a depressing record of destruction, directed not only against the earth he inhabits, but against the life that shares it with him" (Carson 1962, 83).

Sustainable Development

Gro Harlem Brundtland (1939–) Oslo (Norway)

Gro Harlem Brundtland earned both a Master of Public Health (1965) and Doctor in Medicine (1963). She was prime minister of Norway (1981, 1986–89, and 1990–96), and Director of the World Health Organization (WHO) (1998–2003). In the 1980s, she was appointed by the UN to form a commission in order to conduct research into environmental problems and the development model in place at the time. Between 1984 and1987, the commission prepared the so-called Brundtland Report (WCED 1987), in which the term *sustainable development* was coined. In 2000, she was appointed general director of the WHO. In her work, Brundtland has denounced the North–South differences in preventing disease and death related to disease in developing countries, and has shown how industrialized countries can contribute toward ending them:

> The prospects of intervening to prevent death in developing countries have never been better. The evidence refutes those who doubt that the world's poorest communities can be protected from AIDS, tuberculosis (TB), malaria, childhood diseases and maternal mortality. With a concerted effort from the international community we can turn the promise of these success stories into a reality in the coming years (WHO 2000).

In 2007, Brundtland was elected Special Envoy of the General Secretary of the UN on climatic change.

While Rachel Carson, Barbara Ward,[1] and Gro Harlem Brundtland were busy advancing ecological questions on the national and international agenda in the North, other women, such as Wangari Maathai and Vandana Shiva, were doing what Ellen Swallow Richards had imagined one hundred years earlier: expressing their concerns about the ecological degradation of their own communities (Mellor 1997).

The Role of Women in a Vital Change of Direction: Women's Networks

Different environmental movements that have emerged from women's networks show how women have constructed a relationship with the natural world, including the Green Belt movement in Kenya, the Chipko movement in the remote villages of the Himalayas, the Love Channel campaign in the United States, and the women's camp at Greenham Common in Great Britain (Mellor 1997). Two of these movements are analyzed in the sections that follow through the biographies of the women who led them.

The Green Belt Movement

Wangari Maathai (1940–2011) Nyeri (Kenya)

The first woman in central and western Africa to obtain a doctorate (in biology), Wangari Maathai was a professor of veterinary anatomy at the University of Nairobi, where she held virtually all teaching posts, was also the first woman to head a department, and even managed to become the first dean in 1940. An environmental activist, in 1977 she founded the Green Belt Movement, which planted over 20 million trees in an attempt to avoid deforestation and the resultant desertification. The movement concentrates its efforts on environmental conservationism, reforestation, environmental education, Pan-African training workshops on environmental issues aimed specifically at women, and cycles of visits and tours (the so-called Green Belt Safaris). In the main, the program is developed by women who, through receiving payment for their work planting trees, are then better prepared to look after their children and to care for the future of the environment. In 1986, the movement established the Pan-African Green Belt Network with 40 people who replicated the program in their own countries or established similar programs. In 1991, Maathai was arrested, imprisoned, and then freed thanks to a campaign carried out by Amnesty International. In 1999, she was attacked when planting trees in the public forest of Karura in Nairobi in protest against continued

1. Barbara Ward was a British academic who was a pioneer in the critique of Western technology. Ward criticized the adverse effects on the South of the modernizing drive toward global economic development (Ward and Dubos 1972).

deforestation. As co-president of the Jubilee 2000 Africa Campaign, Maathai extended her protests, playing a key role in the demand for the cancellation of the debt of Third World countries. From January 2002, she occupied the role of Visiting Fellow at the Global Institute of Sustainable Forestry at Yale University. She was Deputy Minister of Environment, Natural Resources, and Wildlife (2003–2005), received numerous prizes and mentions for her work, such as the Woman of the Year award in 1983, and was considered by *Earth Times* as one of the 100 people whose work was capable of making a difference in the world in relation to the environment. Maathai was awarded the Nobel Peace Prize in 2004 for her contribution to democracy, peace, and sustainable development.

As M. Dolores Sánchez (2008) notes, the activities carried out by the Green Belt Movement have a dual aim: to conserve the environment and to increase women's power so that they will have a better understanding of their communities. This project aims to empower women, constituting a clear example of community development and environmental protection. As well as making a fundamental contribution to peace, the project offers a global approach to the treatment of a complex problem and endeavors to find solutions within the framework of sustainable development. Maathai's management of the movement and the wise advice she offered women has brought about real changes. Solutions do not "come from outside," but rather from within the communities themselves. For Maathai, global protection of the environment is directly related to ensuring peace:

> The world needs a global ethic with values which give meaning to life experiences . . . Mankind's universal values of love, compassion, solidarity, caring and tolerance should form the basis for this global ethic which should permeate culture, politics, trade, religion and philosophy. It should also permeate the extended family of the United Nations (Maathai 1995).

Maathai was always at the forefront of the fight to promote a form of social, economic, and cultural development that would ultimately prove ecologically viable. Her vision of sustainable development was holistic and encapsulated democracy, human rights, and, in particular, the rights of women. She thought globally and acted locally.

The Chipko Movement

Vandana Shiva (1952–) Dehra Dun, Uttarakhand (India)

Vandana Shiva received a doctorate in quantum physics in 1978 through the University of Western Ontario. A radical ecofeminist, she helped to redefine perceptions of women in developing countries (Shiva, 1988). She is an ecological leader and environmental activist campaigning in the areas of intellectual property, biodiversity, biotechnology, bioethics, and genetic engineering. She cofounded the International Forum on Globalization with Ralph Nader and Jeremy Rifkin. Currently she is Director of the Research Foundation for Science, Technology and Ecology, an independent institute dedicated to researching ecological and social issues.

In 1991, Shiva founded Navdanya, an Indian national movement aimed at protecting the diversity and integrity of living resources, particularly indigenous seeds. In 1993, she received the Right Livelihood prize for putting women and ecology at the heart of contemporary discourses on development, and has also received the Global 500 prize of the United Nations Environment Programme. Shiva is the author of numerous publications of note (Shiva 1991, 1993, 1997, 2000, 2001, 2002; Mies and Shiva 1993), as well as assistant editor of *The Ecologist.*

For Shiva, as an ecofeminist in a Third World country, the exploitation and destruction of the natural world is intrinsic to the dominant industrial economic model of development that she defines as a colonial imposition by the First World. Development that is based exclusively on technological and economic growth has changed relationships between human beings and nature (understood within Indian cosmology as Mother Earth) and locates man above nature, endowing him with the capacity to control and dominate nature, which is considered to be inert and passive. According to Shiva (1988, xvii), "a science that does not respect nature's needs and a development that does not respect people's needs inevitably threatens survival." She concurs with Carolyn Merchant (an American ecofeminist) that women and nature share the experience of being oppressed by patriarchy, and that this shared experience thus defines the connection between them both and establishes this connection at an ideological and material level. Rural women of the Third World obtain 60 to 80 percent of the food they need to live and to feed their families from nature. The destruction of nature thus represents a threat to their lives and the lives of their families.

On the basis of the study carried out by the nonviolent Chipko movement in the state of Uttar Pradesh in the north of India, Shiva explains that women of the Third World have a special dependency on nature and, at the same time, specific knowledge of it. This knowledge, acquired through the accumulation of experiences and their transmission from one generation of women to the next, has been systematically marginalized by modern science and the dominant model of development, both of which are patriarchal constructs that exclude women as experts as well as other types of knowledge (which are more respectful of nature) from knowledge and science.

If we transfer the ecofeminist debate to Third World countries, there is an increasing need to examine the relationship that women establish with the environment, in many cases understood as their very livelihood, at other levels on the margins of ideology. Rural women of Third World countries maintain a close relationship with nature—as users and managers of natural resources, as producers of food and other goods to be consumed or sold, and as administrators and consumers of goods.

Shiva's work has been criticized by various Indian researchers, such as Geeta Menon, Mira Burra, and Bina Agarwal, who present different interpretations of women's relationships with nature. She has also been criticized for attributing the destruction of nature and the oppression of women to colonialism and to the imposition of science and Western development, ignoring the existence of the economic and social inequalities that perpetuated such destruction and oppression prior to colonialism, since it cannot be claimed that the society of pre-colonial India was a fair one or one that was entirely ecologically respectful. In view of the deficiencies presented by ecofeminism and Shiva's position, which is interpreted as the ecofeminist voice of the Third World, an alternative is presented in environmental feminism.

Ecofeminism: The Gender Perspective in Environmental Awareness

Ecofeminism emerged out of the encounter between feminism and ecology. Perhaps it is precisely because of this duality that it still remains a great unknown for the two movements, even though, in its different strains, it opens up a promising horizon for feminists and ecologists.

It is important to highlight that being an ecofeminist does not imply any affirmation that women are innately more linked to nature and to life than men. From a constructivist perspective of gendered subjectivity, the interest that women have according to international studies in ecological matters is not an automatic mechanism related to their sex. As with other aspects of gendered identities, reality shows us great varieties of individuals, as well as tendencies that are linked to socialization in certain tasks and attitudes. The female collective, in general, has not had access to arms and has traditionally been more responsible for the tasks of caring for the more vulnerable members of society (children, the elderly, and the sick), and for maintaining the infrastructure of the household (kitchen, clothes, etc.), developing, in statistical terms, a "relational" subjectivity that is mindful of others and tending to be more emotionally expressive. When these characteristics are combined with being accurately informed and having a healthy distrust of hegemonic discourses, the conditions then become ripe for their interest in ecology to be awakened (Puleo 2008, 47–48).

In 1974, Françoise d'Eaubonne adopted the term *ecofeminism* for the first time, highlighting the historical, biological, and social connections between nature and women and emphasizing that the exploitation and oppression of them both is a consequence of the dominance of man and the patriarchal order (Eaubonne 1974). Ecofeminism considers women as a unique category without making distinctions based on class, cast, race, ethnic groups, or age—a position that it maintained until the 1980s and that upheld a view of matriarchal values and the implementation of a feminine culture.

At the beginning of the 1990s, when this concept was still accepted and continued to grow in popularity, the first theoretical considerations emerged from "environmental feminists," bringing about new lines of thought on the interactions between women, the environment, natural resources, and sustainable development. New tendencies emerged from the original ecofeminism, all of them fueled by concerns and interests in changing the relationships between people and the environment.

Environmental Feminism: A New Ethical and Political Project

Bina Agarwal (1951–) (India)

Feminist economist and an advocate of ecological feminism, Bina Agarwal is professor of the Institute of Economic Growth, University Enclave, New Delhi; is Vice President of the International Economic Association; and was president of the International Association for Feminist Economists. She is Visiting Professor of the Universities of Harvard, Michigan, and Minnesota, and founder of the Indian Society for Ecological Economics. Agarwal has received various prizes, including an honorary doctorate in 2007 and the Padma Shri Award in 2008, for her contribution to the field of literature and education. She has also published numerous books (Agarwal 1986, 1988, 1989, 1994, 2005; Folbre et al. 1991).

In Third World countries, poor rural women depend entirely on the natural environment for their survival and for that of their families and communities. As a result, this relationship is part of their most immediate and palpable reality and not an ideological construct. The connection that these women establish with nature forms the basis of what Agarwal calls *ecological feminism*, an alternative approach to the radical ecofeminism advanced by Shiva. Both Agarwal and Shiva are Indian researchers on feminism and ecology

Ecological feminism is the axis of a debate that emerges from the critiques and proposals of authors from Third World countries and is a perspective that is shared by feminists from both the North and South who are concerned about the environment. Ecological feminism advances the claim that the relationship of women with the environment varies from one woman to another, depending on the social class, race, cast, religion, ethnic group, and so forth to which they belong. An ecological feminist approach explores the factors that determine the effects of environmental degradation on women and their capacity to respond to them. At the same time, these factors also determine the relationship that women have with the organization of production, reproduction, and distribution.

Ecological feminism recognizes that environmental destruction particularly affects women and the poorest populations of Third World countries. However, when analyzing the mechanisms of environmental degradation, ecofeminism attributes part of the responsibility to the dominant groups who monopolize power, property, privileges, and the control of resources. Referring to the Indian experience, Agarwal adds that ecofeminist discourse ignores the real relationship that women establish with nature; this relationship is completely different for each woman and quite different to the interpretation that a person outside the situation may make of it. For Agarwal, from a feminist perspective we must challenge and transform the gender system, the gender-based division of labor, and the unequal distribution of resources according to gender. And from an ecological perspective, we must challenge and change the relationships between people and the environment or nature, as well as put an end to the processes whereby a minority appropriates natural resources—whether these minorities are from developed countries or from the oligarchies of the Southern countries—at the expense of the majority. In view of this, ecological feminism emphasizes the need to fight and transform from feminism and environmentalism at the same time (Llort i Juncadella 1994).

Val (Morrell) (Routley) Plumwood (1939–2008) (Australia)

Beginning in 1960, the Australian ecofeminist and philosopher Val Plumwood was an activist in movements aimed at conserving biodiversity and avoiding deforestation, and helped to establish the interdisciplinary studies known as ecological humanities. When she married philosopher Richard Routley in 1983, they both changed their names—she chose Plumwood and he chose Sulvan. She took her name from the forests of the ancient *Eucryphia moorei* tree, which she actively campaigned to conserve. She worked in the universities of Montana and Sydney and belonged to the Australian Research Council of the Australian National University. In the 1970s and 1980s, Plumwood pioneered the organizing of campaigns to save the rainforest of eastern Australia, extending her fight to the protection of all of the ancient forests of the continent. Her life changed when she was attacked by a crocodile in the National Kakadu Park in northern Australia, and she realized how a human being can go from being predator to becoming prey and just a part of the food chain.

Plumwood criticized what she called "the standpoint of mastery," a set of views of the self and of one's relationships with others that is associated with sexism, racism, capitalism, colonialism, and the domination of nature. She used feminist theories to analyze this standpoint, which includes a view of women as radically separate and inferior. In Plumwood's opinion, Western civilization's view of nature can be understood in the context of a series of problematic dualisms that can be traced back to Greek thought. In this sense, the environmental crisis should be seen as a crisis of reason conceived in opposition to the emotions. The relations of the Western world have been established in accordance with different dualisms: human being/nature; mind/body; reason/emotion; masculine/feminine; civilized/primitive.

The result of these dualisms has been the establishment of the "logic of domination" and the "problem of discontinuity." The logic of domination consists in positing one pole of dualism as superior to the other and thereby attributing the superior pole the right to dominate the inferior pole. In the case of the human/nature dualism, the logic of domination materializes in the modern presumption that nature/material resources should be exploited without limits. Nature, devoid of reason, is thus converted into an instrument for the good of humankind and constitutes the inferior pole of the dualism. The human/nature dualism corresponds to what Plumwood (2002) termed *hyperseparation* and the problem of discontinuity, according to which it becomes extremely difficult to see people as "part of" nature, and as a result of which the dichotomy between human and nature is accentuated further, with the latter acquiring an inferior quality. She argues that homogenization leads to a series of underestimations of the complexity and irreplaceability of nature and explains how, in our androcentric culture, differences are considered as signs of inferiority and represent the basis for the establishment of hierarchies. In view of this, Plumwood postulated that ethical theories based on unity cannot provide a model for any kind of mutual adjustment, communication, or negotiation between the different parties and, thus, do not help to construct ethical relations.

Conclusions

Women have been and are fundamental agents in raising awareness of the ecological crisis and have participated and participate actively in the construction of a more sustainable model of development. The gender perspective continues to be necessary when analyzing factors that influence environmental attitudes and behaviors, bearing in mind that the promotion of such attitudes must aim to involve men more in the private sphere and women more in the public sphere. I agree with Alicia Puleo (2008, 48–49), who states that women are not the saviors of the planet nor the privileged representatives of nature; they can, however, contribute to a sociocultural change toward equality that will ensure that the practices of caring for others—that have historically been located within the female domain—become universalized both in developing countries as well as in advanced industrialized societies. In other words, men and women should take on these practices and they should be extended to the nonhuman, natural world.

References

Agarwal, Bina. 1986. *Cold Hearths and Barren Slopes: The Wood-Fuel Crisis in the Third World.* Riverdale, MD: Riverdale Co.

——. 1994. *A Field of One's Own: Gender and Land Rights in South Asia.* Cambridge and New York: Cambridge University Press.

——. 2005. *Psychology, Rationality and Economic Behaviour: Challenging Standard Assumptions.* Houndmills, Basingstoke, Hampshire, UK and New York: Palgrave Macmillan in Association with International Economic Association.

——. 1988. *Structures of Patriarchy: State, Community and Household in Modernising Asia.* New Delhi: Kali for Women.

——. 1989. *Women, Poverty and Ideology in Asia: Contradictory Pressures, Uneasy Resolutions.* Basingstoke, UK: Macmillan.

Agarwal, Bina, Jane Humphries, and Ingrid Robeyns, eds. *Capabilities, Freedom and Equality: Amartya Sen's Work from a Gender Perspective.* New Delhi and New York: Oxford University Press.

Alvarez-Lires, Mari, Teresa Nuño, and Nuria Solsona. 2003. *Las científicas y su historia en el aula.* Madrid: Síntesis.

Alvarez-Lires, Mari. 2008. "Género y desarrollo sostenible: Génesis del concepto desarrollo sostenible." Unpublished. Development aid course material at the Universities of Oriente and Havana, Santiago de Cuba.

Boserup, Ester. 1970. *Woman's Role in Economic Development.* New York: St. Martin's Press.

Carson, Rachel. 1955. *The Edge of the Sea.* Boston: Houghton Mifflin.

——. 1951. *The Sea Around Us.* New York: Oxford University Press.

——. 1965. *The Sense of Wonder.* New York: Harper & Row.

——. 1962. *Silent Spring.* Boston: Houghton Mifflin Co.

——. 1941. *Under the Sea Wind: A Naturalist's Picture of Ocean Life.* New York: Simon and Schuster.

Eaubonne, Françoise d'. 1974. *Le féminisme ou la mort.* Paris: P. Horay.

Folbre, Nancy, Barbara Bergmann, Bina Agarwal, and Maria Floro, eds. 1991. *Women and Work in the World Economy.* London: Macmillan.

Llort i Juncadella, Imma. 1994. *Ecofeminismo(s) o Feminismo Ecologista.* At

http://sexualpersonae.wordpress.com/2008/06/12/feminismo-ecologico-desde-el-blog-callejera-bina-agalwal/.

Maathai, Wangari. 1995. "Bottle-necks of Development in Africa." Speech at the Fourth United Nations World Women's Conference, Beijing, China, August 30–September 15. At http://gos.sbc.edu/m/maathai.html.

Margulis, Lynn. 2002. *Acquiring Genomes: A Theory of the Origin of Species*. New York: Basic Books.

——. 1981. *Symbiosis in Cell Evolution: Life and Its Environment on the Early Earth*. San Francisco: W.H. Freeman.

——. 1998. *Symbiotic Planet: A New Look at Evolution*. New York: Basic Books.

Margulis, Lynn, and Dorion Sagan. 1986. *Microcosmos: Four Billion Years of Evolution from Our Microbial Ancestors*. New York: Summit Books.

Margulis, Lynn, and Karlene V. Schwartz. 1982. *Five Kingdoms: An Illustrated Guide to the Phyla of Life on Earth*. San Francisco: W.H. Freeman.

Mellor, Mary. 1997. *Feminism and Ecology*. Cambridge: Polity Press; Washington Square, NY: New York University Press.

Mies, Maria, and Vandana Shiva. 1993. *Ecofeminism*. Halifax, NS: Fernwood; London: Zed Books.

Plumwood, Val. 2002. *Environmental Culture: The Ecological Crisis of Reason*. London and New York: Routledge.

Puleo, Alicia H. 2008. "Libertad, igualdad, sostenibilidad: Por un ecofeminismo ilustrado." *Isegoría. Revista de Filosofía Moral y Política* 38 (January–June): 39–59.

Richards, Ellen H. 1900. *Air, Water, and Food from a Sanitary Standpoint*. New York: Wiley.

——. 1911. *Conservation by Sanitation: Air and Water Supply; Disposal of Waste*. New York: Wiley.

——. 1886. *Food Materials and Their Adulterations*. Boston: Estes and Lauriat.

——. 1896. *Laboratory Notes: Sanitary Chemistry and Water Analysis*. N.p.: N.p.

Sánchez, M. Dolores. 2008. "Género y desarrollo sostenible: Las redes de mujeres." Unpublished. Development aid course material at the Universities of Oriente and Havana, Santiago de Cuba.

Shiva, Vandana. 1997. *Biopiracy: The Plunder of Nature and Knowledge*. Toronto: Between the Lines.

——. 1993. *Monocultures of the Mind: Perspectives on Biodiversity and Biotechnology*. London and Atlantic Highlands, NJ: Zed Books; Penang, Malaysia: Third World Network.

——. 2001. *Protect or Plunder? Understanding Intellectual Property Rights*. London and New York: Zed Books.

——. 1988. *Staying Alive: Women, Ecology and Survival in India*. London: Zed Books.

——. 2000. *Stolen Harvest: The Hijacking of the Global Food Supply*. Cambridge, MA: South End Press.

———. 1991. *The Violence of the Green Revolution: Third World Agriculture, Ecology, and Politics.* London and Atlantic Highlands, NJ: Zed Books.

———. 2002. *Water Wars: Privatization, Pollution and Profit.* Toronto: Between the Lines.

Ward, Barbara, and René Dubos. 1972. *Only One Earth: The Care and Maintenance of a Small Planet.* New York: Norton.

WCED (World Commission on the Environment and Development). 1987. *Our Common Future: Report of the World Commission on the Environment and Development* [Brundtland Report]. General Assembly, United Nations, Forty-second Session, Supplement no. 25, A/42/25.

WHO (World Health Organization). 2000. "Poorer Half of the World Can Expect Better Health and Prosperity Within the Next Decade." Press Release WHO/78, December 19. At www.who.int/inf-pr-2000/en/pr2000-78.html.

11

Environmental Education as Training: A Case Study at the University of the Basque Country (UPV/EHU)

Araitz Uskola Ibarluzea
The University of the Basque Country (UPV/EHU)

The five objectives of Environmental Education (EE) formulated by UNESCO (1977) range from increasing awareness of and knowledge about environmental problems to making the leap to action and social participation. Nevertheless, during the 1990s, it was noted that EE programs emphasized progress toward those objectives related to knowledge acquisition and raising awareness, forgetting those related to competence-based training and citizen participation in managing environmental problems.

Practical experience of the complexity of both natural systems and environmental problems has led to modifications in educational approaches, both in EE and in other areas, such as science teaching (OECD 2007), with an emphasis on the aim of developing well-informed students who are capable of addressing socio-scientific problems, making decisions, and taking action accordingly: in sum, educating for citizen participation.

With a similar understanding of science, Silvio O. Funtowicz and Jerome Ravetz (1993) have introduced the concept of "postnormal" science and believe that in the case of socio-environmental systems (which entail high levels of risk and uncertainty), individuals' values are fundamental, not secondary, elements. In this way, both experts and nonexperts acquire a more balanced power of decision making, given that there are no definitive solutions endorsed by science.

Thinking about complexity has likewise been reflected in the need to change educational proposals in EE through institutional documents and declarations. For example, the United Nations (UN) Decade of Education for Sustainable Development 2005–2014 program (UNO 2002) "breaks down the traditional educational scheme and promotes a host of different approaches: interdisciplinary and holistic learning rather than subject-based learning;

values-based learning; critical thinking rather than memorizing; multi-method approaches: word, art, drama, debate, etc.; participatory decision-making; [and] locally relevant information, rather than national" (UNESCO INRULED 2004–2010).

Environmental Education as Training for Action

In the context of EE, several Danish environmental educators (Breiting 1995; Jensen and Schnack 1997) have called for a new EE that is directed toward *training for action* and that takes into account the need to consider the conflicts of interest entailed by every environmental problem. In this way, the abilities to think critically, to clarify one's own values, to put oneself in somebody else's shoes, to discern the data on which an argument is based, to make a decision, and to take action in consequence become the fundamental educational objectives to be pursued.

Bjarne Bruun Jensen and Karsten Schnack (1997) talk about action competence, associating competence with the ability and desire to be a qualified participant, and emphasizing the intentionality of actions in order to distinguish them from behaviors, activities, and habits. They contrast the search for action competence to the search for behavioral change, a primary objective of the majority of current EE activities and programs. In behavior-changing activities, educators decide what is the best behavior for the good of the environment, basing their decision on the certainty of scientific analyses, without taking into account other factors, such as the values of the individuals involved. If the activity is successful, the participants will demonstrate appropriate behavior, but it is unlikely that they will have developed the competence to take action in response to new problems or to jointly construct a sustainable society.

In activities designed to train for action, in contrast, the educational process is more important than the product. From this perspective, the educator should not direct his or her efforts toward achieving a specific change in behavior, but rather should facilitate scenarios that can develop students' abilities so that they can decide, using their critical skills and in a democratic way, what direction change should take. The aim, consequently, is to develop "a critical, reflective and participatory approach in which the future adult can cope with environmental problems in a democratic way," instead of "prescribing to pupils certain behavioural patterns here and now that we believe contribute to solving current environmental problems" (Mogensen and Mayer 2005, 14).

This new perspective entails a major change when it comes time to formulate the objectives of educational interventions, since an activity may have high educational value despite not having high environmental value. These activities will have to be evaluated "on the basis of their educational value and thus according to educational criteria. . . . The crucial factor must be what the students learn from participating in such activities, or from deciding something else" (Jensen and Schnack 1997, 165).

Decision-Making Ability

From the perspective of training for action, the participants are the ones who, using their critical skills, decide the direction of the actions to be taken. The ability to make decisions is, therefore, considered one of the fundamental abilities that EE should develop. Various aspects of this ability have been studied: the number and kind of criteria used to make a decision, the consideration or lack thereof of "conflicting" criteria, the priorities among criteria, and whether the process is compatible with a normative decision-making framework. In the case of environmental problems, it becomes important to analyze the types of criteria constructed and used in decision making, paying special attention to the category of ecological criteria (Hogan 2002; Jiménez-Aleixandre and Pereiro-Muñoz 2002; Patronis, Potari, and Spiliotopoulou 1999; Wu and Tsai 2007).

One way to analyze a decision-making process is to compare it to normative frameworks for that process (Kortland 1996; Ratcliffe 1997), according to which someone first accepts the challenge of deciding, then looks at what the alternatives are, establishes criteria, evaluates the alternatives using those criteria, makes a decision, and analyzes how to implement it (Janis and Mann 1977). Studies have concluded that the process in daily life usually does not follow these steps exactly, since formulating alternatives already includes evaluating them to some extent, and sometimes the decision is even made at the beginning, conditioning the subsequent process in such a way as to take into account only the information that supports the decision.

Along these lines, Koos Kortland (1996) studied the decision-making process of thirteen- and fourteen-year-old students who were making a decision about buying milk in different containers, a subject related to waste management. He concluded that the students considered few criteria and those they did consider favored their choice, so that they avoided prioritizing conflicting criteria. In a related analysis, Mary Ratcliffe (1997) studied the processes used by fifteen-year-old students in the course of five decision-making tasks. She developed a six-step normative framework that she had her students explicitly follow when she presented them with the situation about which they had to decide. She analyzed the general procedure followed by the groups, studying the sequence of steps and measuring the time dedicated to each one. She concluded that her students began to evaluate options at the same time that they identified criteria. In the cases in which the students were capable of explicitly identifying important criteria, this helped to focus their discussions, since they were more attentive to information relevant to those criteria and evaluated the options more systematically.

Developing Critical Thinking

Developing critical thinking is, as the UN proposes (UNO 2002), one of the objectives that EE should pursue. Finn Mogensen and Michela Mayer (2005) even consider this ability the chief prerequisite for the development of action competence.

Several authors identify critical thinking with the ability to evaluate the consistency of a statement supported by evidence (Siegel 1988; Kuhn 1991). Studies on argumentation are acquiring special relevance in the field of research into the teaching and learning of science (Erduran and Jiménez-Aleixandre 2008), something of great interest for EE.

For some authors, on the other hand, the highest level of this ability is manifested in the ability to think through positions opposed to one's own and the evidence that supports them, as well as to rebut others' arguments by using appropriate evidence (Erduran, Simon, and Osborne 2004; Kuhn 1991; Zohar and Nemet 2002). In this way, rebuttal is understood as a need to question the data for the opposing arguments. Sibel Erduran, Shirley Simon, and Jonathan Osborne (2004) developed an instrument to measure the quality of argumentation on the basis of the presence of rebuttals in the arguments put forward in episodes of clashing opinions. María Pilar Jiménez-Aleixandre, Ramón López Rodríguez, and Erduran (2005) modified that instrument and introduced one with five levels of quality.

Nevertheless, Jiménez-Aleixandre and Blanca Puig (2010) believe that this component related to argumentation has to be supplemented by one related to social emancipation. They propose that critical thinking also includes the ability to express an independent opinion and the ability to critically analyze discourses that justify inequalities. The inclusion of skills of this kind under the heading of critical thinking is based on critical theory questioning the higher value placed on technology than on democracy in current society and advocating for a transformation of reality and power relations. They consequently define critical thinking as "the competence to develop independent opinions and to develop the ability of reflecting about the world around us and of participating in it" (Jiménez-Aleixandre and Puig 2010, 1005). This definition is fully consonant with the transformations required by a transition to a more sustainable society, for which reason it will be the one that we will adopt when referring to critical thinking as a capacity to be developed by EE.

Contexts for the Development of Competence-Based Training and Critical Thinking

The development of decision-making and critical-thinking skills and reflection on the complexity of environmental problems requires complex contexts of teaching and learning. Along these lines, some authors (Colucci-Gray et al. 2006, 248) propose "to integrate the more traditional ways of teaching science . . . with direct experience of complexity, as for example: the study of complex issues; the use of complex teaching methods (simulations, group discussions, debates); the involvement of different levels of personal development." Jiménez-Aleixandre (2008) urges tackling authentic problems, those that do not have an immediate or obvious solution; are relevant to the students, because they are situated in a real-life context; require investigation to solve; and are open to different responses or different processes.

The debates generated in the discussion of such problems constitute a learning context that offers opportunities to develop argumentative competencies, since it puts students in the position of justifying their affirmations. If the problem also includes a social component, and scientific data is not sufficient to resolve it, it will likewise provide an opportunity to develop those aspects of critical thinking associated with social emancipation (Jiménez-Aleixandre and Puig 2010).

Problems of this kind also constitute an opportunity for knowledge construction (Sadler, Barab, and Scott 2007), since they form a context within which the content treated makes sense to the students. Improved conceptual understanding following a task including debate has been shown statistically by Fang-Ying Yang (2004), who analyzed the number of directly and indirectly connected ideas in conceptual maps extracted from student discourse before and after participating in the task. Nevertheless, Kortland (1997) did not observe such an improvement and notes that even if students learn from the proposed task, they do not learn as much as desired, and the task itself does not guarantee that a conceptual conflict will result. This may agree with the findings of Claudia von Aufschnaiter et al. (2008), who put forward the thesis that it does not appear that the process of debate causes students to construct new knowledge, but that it constitutes an excellent opportunity for them to develop a deeper understanding and firmer grasp of their ideas, helping them to make connections across contexts and thereby enabling better subsequent learning.

Educational interventions in EE are directed toward mobilizing individuals' values, since those values will turn out to be key when the time comes to make decisions and take action. As a consequence, cooperative work in groups that try to solve an open environmental problem may constitute a strategy that facilitates the acquisition of values characteristic of a sustainable society.

Nevertheless, when cooperative tasks are proposed, group dynamics are not always egalitarian, and what happens in the group depends to a great extent on those dynamics. Along these lines, several authors (Anderson 2007; Oliveira and Sadler 2008) believe that both social and cognitive processes are responsible for knowledge construction, a view that combines cognitive and sociocultural perspectives. Virginie Albe (2008) identified three categories in the oral discussions that took place in two groups of six students each: acceptance, collaborative argumentation, and contradictory confrontations. Acceptance consists of accepting the proposals of one member without debate. Collaborative argumentation may take place when knowledge is constructed on a collaborative basis or when different opinions are explicitly present within a group and this leads to the need to justify them. Disagreements, on the other hand, may give rise to contradictory confrontations in which the opinions are not justified, with the result that the object of discussion is destabilized and sometimes even abandoned. Alandeom W. Oliveira and Troy D. Sadler (2008) observed dynamics similar to those found by Albe (2008) in the three groups they analyzed, and they concluded that only the one in which a positive atmosphere was created was able to make conceptual progress. Lucia Mason (1996) agrees that both co-construction and critical opposition make it possible to negotiate and share ideas, favoring the construction of concepts.

Training for Action with UPV/EHU University Students

The following lines detail some of the results obtained in an educational experience designed according to the framework previously described. The intervention took place in a section of the elective course Environmental Education and Non-Formal Education in the degree program in social education at the UPV/EHU. Twenty-five students (twenty-three female and two male), approximately eighteen to twenty-three years old and divided into four groups, were instructed to decide on a heating system for a future building, choosing among diesel, natural gas, propane, biomass, and electricity. They were given a twenty-three-page dossier of information on advantages and disadvantages of the different energy sources and were provided several articles. They were not told what criteria they should use to make a decision.

Three sessions were dedicated to cooperative work in groups to read the information and make a decision and were captured on audio recordings for later analysis; the fourth session was dedicated to sharing the results and was recorded on video. The findings presented are based on the analysis of the oral production of group A (made up of six young women) during the first three sessions. The young women's real names have been replaced by fictitious ones. This study focuses on giving a description of this group due to the fact that it stood out in several interesting aspects that it is desirable to highlight here.

A summary of the most noteworthy findings with regard to the development of decision-making ability and critical thinking follows. This paper aims to understand the phenomena studied through the meanings ascribed by actors to their actions (Erickson 1986). The author is the professor who taught the course and acted as a participant observer during the sessions in which the activity was carried out. Of the mentioned aspects related to decision-making ability and critical thinking, the following will be addressed:

- The decision-making process and its compatibility with normative reference frameworks
- The selection and use of criteria on which to base the decision and the evaluation of those proposed
- The evaluation of knowledge on the basis of evidence
- The ability to rebut opposing positions on the basis of evidence
- The questioning of authority
- The ability to keep an open mind
- The consideration of social inequalities and inequalities of power

Steps in the Decision-Making Process

In order to analyze the transcriptions of the discussions among the students, a four-step framework was selected:

1. *Proposing criteria* (PC): constructing or identifying the criteria that are going to be taken into account in the process
2. *Searching for information* (SI): seeking and clarifying information about the alternatives, taking into account the criteria considered important
3. *Evaluating the alternatives* (EA): assessing the alternatives in accordance with the criteria proposed
4. *Making a decision* (MD): deciding what option is best in accordance with the analysis conducted

The students' 1,001 interventions during the first two discussion sessions were accordingly classified in terms of these steps in the decision-making process, with the result that 75 percent of the total can be considered as belonging to that process. In terms of the time dedicated by group A to the different steps in the decision-making process, the data indicate that 49 percent of all the oral interventions were dedicated to evaluating the alternatives (of the 1,001 interventions: PC: 1 percent; SI: 20 percent; EA: 49 percent; MD: 5 percent).

At the beginning, without information, all of the students in group A opted for natural gas and justified that decision on the basis of their view of it as ecologically sound, cheap, and convenient. Nevertheless, they were aware that they lacked information in order to be able to make a reasoned decision and that such a decision had to be based on an evaluation of the advantages and disadvantages of the different alternatives.

After reading the information, Amaia became aware of the finite nature of fossil fuels, and that led them to drop their first choice, after which they explicitly formulated some criteria to take into account in making a decision. At that point, they appeared to be acting "correctly" according to normative decision-making frameworks, putting the criteria that were going to govern the process on the table before rushing to commit to one of the options. In the following excerpt from the transcription (table 11.1), we can see the variety of criteria used by the members of this group when it came time to evaluate the alternatives, as well as the formulation of criteria to be taken into account in the process.

Table 11.1. Excerpt from the transcription of interventions 164–179 of group A (EA: Evaluating alternatives; PC: Proposing criteria).

Line	Student	Transcription	Step in the decision-making process
165	Ane	[In reference to fossil fuels] I think that I'm not finding any advantages.	EA
167	Arantza	Man, the stuff in the cylinders, no, but natural gas, that comes to your house.	EA
168	Amaia	As far as convenience for living.	EA
169	Unidentified	And the cost?	PC
172	Ainara	The cost is low, the cost for natural gas.	EA
173	Amaia	Man, well, that also, that also interests us, egotistically.	EA
176	Amaia	You look for what causes the least harm and what benefits you the most.	PC
177	Arrate	A mix of everything, the cheapest that causes the least harm.	PC
178	Ainara	And another disadvantage is that it's not renewable, that it's gone, and . . .	EA
179	Ane	That if it's gone, it's gone.	EA

In this phase, they did not make decisions. The subsequent decision was to rule out electricity (intervention no. 268), but that position led two students to propose opting for electricity and installing solar panels on the building, with the result that there was a difference of opinion within the group from that point on.

In the second session, after each student had explicitly stated what her position was and why, there was an intervention that can be considered key for the subsequent process. This was intervention no. 533, in which Arrate made explicit the criterion of taking future generations into account in order to rebut Ainara, who had noted that "there are still years left" as a warrant in favor of natural gas (see table 11.2).

On the dialectical level, a notable change took place. In intervention 589, instead of engaging in rebuttal (as she had been doing), Amaia set out conditions for a shift from supporting the gas option to supporting electricity, initiating a process of negotiation in which Arantza (591, 594) appeared to make concessions ("it doesn't come with labels," "it's expensive") in order to convince:

> 594 Arantza: In other words, [electricity produced using solar panels] is indeed going to be more expensive, but that expense . . . As you go into the future, it's going to be the only one that's going to be left. It's that, look, it says here that petroleum [will only last] until 2045.

Before each rebuttal, Arantza first made a concession that consisted of disarming the previous arguments by first making statements of concession and then weakening or attacking those statements. In the course of the process, three people were convinced.

As far as the decision-making process followed is concerned, it is difficult to say whether it fit within the normative frameworks considered or not, because the process followed was more complex. The normative framework used by Kortland (1996) and Ratcliffe (1997) is more of a linear one and may be valid for an individual. In this case, however, group dynamics and the requirement of the task to reach a group decision led one individual to construct a criterion after adopting a position, something that in principle is not compatible with the normative framework. In addition, the construction of this criterion led three individuals to adopt a new decision, something that would be compatible with the normative framework for those individuals at the individual level. This sequence of events—which is open to various interpretations at the individual level—is considered to constitute something positive at the level of the group, since it was the group that created a criterion and was capable of adopting it as an important criterion that would guide the decision-making process from that point on.

Criteria Considered

All interventions were taken into account in which a criterion to be applied in the process was being made explicit or was being used in the process, something that constituted implicit evidence of the criterion's importance. This is similar to what Ratcliffe (1997) did and may be the origin of the wide variety of criteria identified.

As far as categorization is concerned, this study proposes two categories for the criteria that make up the "ecological" category in other studies (Jiménez-Aleixandre and Pereiro-Múñoz 2002; Patronis, Potari, and Spiliotopoulou 1999; Simonneaux 2001; Wu and Tsai 2007), dividing them into those that have to do with environmental degradation on the one hand ("pollution") and with resource depletion on the other ("resources"). In addition, sustainability and the future were considered as a distinct criterion ("sustainability"). As a consequence, the kinds of criteria were grouped into seven categories: economic, pollution, pragmatism, resources, convenience, sustainability, and other.

Group A made a total of 490 allusions to criteria, making it possible to conclude that the constant debates between parties defending different positions promoted continual reference to different criteria. Among the criteria, the "economic" criterion and that of "pollution" were most prominent, with 168 and 96 allusions, respectively. The explanation for this prominence lies, on the one hand, in the fact that these are the criteria most commonly debated when dealing with environmental problems, and on the other hand, in the fact that the two opposing sides in this group did not agree on which option had to be considered the cheapest and which the least polluting and debated this until the end. In any event, even if they did not have the same quantitative importance, the "resources" and "sustainability" criteria were key for one party within the group, and it is worth pointing out the ability group A demonstrated in constructing a criterion of its own, that of "sustainability," which was not mentioned in the information dossier provided.

Evaluation of Knowledge on the Basis of Evidence

In argumentation studies, the structure of the argumentative process is analyzed in a variety of ways (see, for example, the review conducted by Sampson and Clark [2008]). In this case, one of those most often used has been adopted, based on identifying the components of arguments as classified by Stephen E. Toulmin (1958). In this perspective, the primary elements are the claims, the data on which they are based, and the warrants that serve as the link between data and claims. There are also other components, such as the modal qualifiers that nuance the claim and the backing that serves as theoretical support for the warrants.

The discussion among the students was divided into episodes, taking the approach that a new episode began when there was a change in the activity being performed or the topic being debated. The arguments constructed by the group in each episode and their component elements were identified. In table 11.2 we see an excerpt from the transcription of episode nine of the second session. This was the episode in which the students, after having debated without being able to reach agreement, took turns giving their opinions and justifying them. In this excerpt, the majority of the elements correspond to argument one, the claim of which is "yes to natural gas" as the preferred option.

Table 11.2. Excerpt from the transcription of interventions 532–535 of group A, session two, episode nine ("Presentation of individual opinions"). (Cx: claim of argument x; Wx.y: warrant no. y of argument x; MQx.y: modal qualifier no. y of argument x; Rb: rebuts)

Line	Student	Transcription	Interpretation
532	Ainara	(. . .) I'm also leaning toward natural gas (. . .). So natural gas is more economical, more convenient, and, well, even if there are few resources, well, screw it, there are still years left, aren't there? And that, well, I don't know.	C1 W1.3 and W1.4 MQ1.2 W1.6
533	Arrate	The thing is, you shouldn't only think about the years you're going to be here; think about those who are going to come after.	W6.2 / Rb W1.6
534	Ainara	Yes, but . . .	
535	Amaia	Sustainable development.	Meaning W6.2

Group A had a high level of ability to evaluate the claims made in accordance with data and warrants (Maguregi, Uskola, and Jiménez 2009), since the group members constructed a total of 99 arguments, 85 of them justified. In addition, they constructed complex argumentative structures with arguments that acted as warrants for other arguments. About 60 percent of the arguments had more than one warrant, with some arguments having up to seven different warrants. As far as the use of evidence or data is concerned, 69 percent of the claims were supported by data or evidence.

Rebuttal of Opposing Positions on the Basis of Evidence

The students in group A were able to rebut opinions opposed to their own by using evidence and not a mere clash of opinions. In the case of group A, of the 73 arguments constructed in episodes of clashing opinions, more than half (50.7 percent) were classified in the two highest levels, those corresponding to rebuttals connected to data or the opponent's warrants. As an example, we see in table 11.2 that Arrate (line 533) rebutted Ainara's warrant W1.6 in support of the choice for natural gas by invalidating it as a warrant. Arrate was not merely saying that natural gas did not seem like a good option to her, but was rather attacking the use of "the fact that there are still [reserves for] years of natural gas left" as a warrant, since under this scenario they would not be available for future generations, something that ("think about those who are going to come after") for her constituted a warrant for argument six, the claim of which is "no to natural gas."

Questioning Authority

Questioning authority is closely related to the previous aspect, since making a value judgment on the basis of evidence or data means acting rationally and not deferring to an opinion on the basis of the authority of its holder. In the case studied, the students did not have an expert opinion to confront, with the consequence that, in principle, they had few opportunities to question authority directly. Nevertheless, they did on occasion demonstrate this ability by questioning the data itself, the dossier they were provided (1090 Amaia: "Yes, but maybe . . . Don't always trust notes.").

An Open Mind

Keeping an open mind is considered another aspect of critical thinking. From the beginning, the students in group A showed themselves ready to evaluate all options and change their opinions if the data supported a better option. This is clear, for example, in the following excerpt from the transcription:

> 93 Ane: So then what? We've already changed our opinion, haven't we?
> 95 (Unidentified): [natural gas] It's not so good now.
> 96 Amaia: Even so, it's better to read them all.
> 97 Arantza: Yes, before drawing conclusions, read everything.
> 98 Amaia: Yes, because after reading natural gas, you think it's bad, and after that you read the rest, and they're worse.

Throughout the process, they were able to realize and acknowledge disadvantages of their own choice on as many as 40 occasions (Uskola, Maguregi, and Jiménez-Aleixandre 2010).

Open-mindedness was most evident in group A in the changes of opinion by several of its members previously mentioned. It is unusual for people to change their attitudes, since we tend to pay attention to information that confirms our interests (Kortland 1996). Nevertheless, Jiménez-Aleixandre and Cristina Pereiro-Muñoz (2005) and Laurence Simonneaux (2001) found changes in attitude and evolving views among students who debated socio-scientific problems. Some of the students studied by Simonneaux (2001) changed their position in the course of a debate (4 of 17) and a role-playing exercise (3 of 17) about setting up a fish farm stocking genetically modified salmon. Simmoneaux indicates that he was surprised by that fact and puts forward the hypothesis that it was the task itself, oral discussion, that was the facilitator of change, since similar results had not been found in previous research. He does not, however, mention possible reasons for the different students to change their opinions.

Separately, out of a total of 37 students who analyzed the construction of a sewer in a Galician marsh (Jiménez-Aleixandre and Pereiro-Muñoz 2005), 16 changed their opinion, and 6 evolved in the direction of a more nuanced position. The causes they identified for this change were the introduction of new data and the arguments advanced in the debate among experts. In this case, the change in attitude produced in this process was caused by the introduction of a criterion by a student during the process of discussion. This shows the high degree of responsibility taken on by this group in the assigned task, as well as the open minds demonstrated by the students who changed their opinions upon considering an unforeseen criterion and one that in principle did not support the chosen option.

Consideration of Inequalities

It was not possible to find many allusions to social inequalities or inequalities of power. Nevertheless, the students were aware that the current rhythm of consumption affects future generations, and some interventions occurred in which intergenerational solidarity is evident. This can also be interpreted as an inequality of power, if not between social classes, still between generations:

> 90 Ane: No, if we're eating all the energy for future generations.
>
> 91 Arantza: Yes.
>
> 92 Amaia: How selfish of us!
>
> 533 Arrate: The thing is, you shouldn't only think about the years you're going to be here; think about those who are going to come after.

They were also aware that resource scarcity will create inequalities among social classes:

> 596 Arantza: Natural gas will last a little longer, but the thing is, inevitably, as less natural gas is left, so the price is also going to go up, and only people who have a lot of money are going to be able to get it. In other words, as the reserves are increasingly depleted, it's going to go up more, what's happening with gasoline; in the end, only rich people are going to be able to have a gas car.

Conclusions

This study has confirmed that university students in the group observed, faced with an open environmental problem, were capable of following a high-quality decision-making process by creating their own criteria, evaluating different options according to those criteria, establishing priorities, and making justified and well-argued decisions. At the same time, the context designed favored the development of different aspects of critical thinking.

Definitively, tasks and methodological proposals of this kind constitute a good framework for working on training in decision making and the development of critical thinking. In addition, in the course of the process, the students in this group were able to jointly construct the meaning of several difficult-to-interpret environmental concepts key to thinking about environmental problems (Uskola, Maguregi, and Jiménez-Aleixandre 2010).

Nevertheless, the task in itself did not ensure the development of the mentioned capabilities or the construction of knowledge. This study has confirmed that the dynamics produced within groups led to very different trajectories within those groups, conditioning both the decision-making process (Uskola, Maguregi, and Jiménez-Aleixandre 2011) and the construction of meaning for environmental concepts (Uskola, Maguregi, and Jiménez-Aleixandre 2010), as well as argumentative ability (Maguregi, Uskola, and Jiménez 2009). In fact, the disagreement that occurred between two parts of the group was key to group A's process, since this disagreement led the group members to justify their positions, allude to multiple criteria, rebut the opposing side's arguments, create a new criterion, and use the knowledge they had available. Other authors have also observed the positive effect of disagreement within groups (Mason 1996; Oliveira and Sadler 2008; Ratcliffe 1997).

As a consequence, there are some variables that we have to take into account and plan for in our intervention, including group management, since we should try to ensure that groups are heterogeneous and have internal dynamics that include a climate of trust that can promote debate (Oliveira and Sadler 2008). Despite these limitations, learning contexts of this kind promote the achievement of EE objectives, for which reason we hope that they can help to train students for action. In other words, as Mogensen and Mayer (2005) note, even if only over the long term, they can help students to participate actively and critically in constructing a more sustainable society. We do have to be aware, of course, that we face "the dilemma of environmental educators, who as environmentalists are in a hurry and as educators cannot be" (Calvo and Gutiérrez 2007, 47).

References

Albe, Virginie. 2008. "When Scientific Knowledge, Daily Life Experience, Epistemological and Social Considerations Intersect: Students' Argumentation in Group Discussions on a Socio-scientific Issue." *Research in Science Education* 38, no.1 (2008): 67–90.

Anderson, Charles W. 2007. "Perspectives on Science Learning." In *Handbook of Research on Science Education,* edited by Sandra K. Abell and Norman G. Lederman. New York: Routledge.

Breiting, Søren. 1995. "Towards a New Concept of Environmental Education." In *Proceedings of the Conference on the Exchange of Promising Experiences in Environmental Education in Great Britain and the Nordic Countries: November 11–13, Karlslunde, Denmark.* Bradford, UK: European Research and Training Centre on Environmental Education, University of Bradford.

Calvo, Susana, and José Gutiérrez. 2007. *El espejismo de la educación ambiental.* Madrid: Morata.

Colucci-Gray, Laura, Elena Camino, Giuseppe Barbiero, and Donald Gray. 2006. "From Scientific Literacy to Sustainability Literacy: An Ecological Framework for Education." *Science Education* 90, no. 2: 227–52.

Erduran, Sibel, and María Pilar Jiménez-Aleixandre, eds. 2008. *Argumentation in Science Education: Perspectives from Classroom-based Research.* Dordrecht: Springer.

Erduran, Sibel, Shirley Simon, and Jonathan Osborne. 2004. "TAPping into Argumentation: Developments in the Application of Toulmin's Argument Pattern for Studying Science Discourse." *Science Education* 88, no. 6: 915–33.

Erickson, Frederick. 1986. "Qualitative Methods in Research on Teaching." In *Handbook of Research on Teaching,* edited by Merlin C. Wittrock. New York: Macmillan.

Funtowicz, Silvio O., and Jerome Ravetz. 1993. "Science for the Post-normal Age." *Futures* 25, no. 7: 739–55.

Hogan, Kathleen. 2002. "Small Groups' Ecological Reasoning while Making an Environmental Management Decision." *Journal of Research in Science Teaching* 39, no. 4: 341–68.

Janis, Irving L., and Leon Mann. 1977. *Decision Making: A Psychological Analysis of Conflict, Choice, and Commitment.* New York: The Free Press.

Jensen, Bjarne Bruun, and Karsten Schnack. 1997. "The Action Competence Approach in Environmental Education." *Environmental Education Research* 3, no. 2: 163–78.

Jiménez-Aleixandre, María Pilar. 2008. "Designing Argumentation Learning Environments." In *Argumentation in Science Education: Perspectives from Classroom-based Research,* edited by Sibel Erduran and María Pilar Jiménez-Aleixandre. Dordrecht: Springer.

Jiménez-Aleixandre, María Pilar, Ramón López Rodríguez, and Sibel Erduran. 2005. "Argumentative Quality and Intellectual Ecology: A Case Study in

Primary School." Paper presented at the NARST (National Association for Research in Science Teaching) Annual Meeting, Dallas, April 4–7.

Jiménez-Aleixandre, María Pilar, and Cristina Pereiro-Muñoz. 2002. "Knowledge Producers or Knowledge Consumers? Argumentation and Decision Making about Environmental Management." *International Journal of Science Education* 24, no. 11: 1171–90.

——. 2005. "Argument Construction and Change while Working on a Real Environment Problem." In *Research and the Quality of Science Education*, edited by Kerst Boersma, Martin Goedhart, Onno de Jong, and Harrie Eijkelhof. Dordrecht: Springer.

Jiménez-Aleixandre, María Pilar, and Blanca Puig. 2010. "Argumentation, Evidence Evaluation and Critical Thinking." In *Second International Handbook for Science Education*, edited by Barry J. Frasser, Kenneth Tobin, and Campbell McRobbie. Dordrecht: Springer.

Kortland, Koos. 1996. "An STS Case Study about Students' Decision Making on the Waste Issue." *Science Education* 80, no. 6: 673–89.

——. 1997. "Garbage: Dumping, Burning and Reusing/Recycling; Students' Perception of the Waste Issue." *International Journal of Science Education* 19, no. 1: 65–77.

Kuhn, Deanna. 1991. *The Skills of Argument*. Cambridge: Cambridge University Press.

Maguregi, Gurutze, Araitz Uskola, and María Jiménez-Aleixandre. 2009. "La competencia argumentativa en la toma de decisiones ante un problema ambiental." *Enseñanza de las Ciencias*, número extra VIII, Congreso Internacional sobre Investigación en Didáctica de las Ciencias, Barcelona: 1098–1101.

Mason, Lucia. 1996. "An Analysis of Children's Construction of New Knowledge through Their Use of Reasoning and Arguing in Classroom Discussions." *Qualitative Studies in Education* 9, no. 4: 411–33.

Mogensen, Finn, and Michela Mayer. 2005. "Perspectives on Environmental Education—A Critical Framework." In *Eco-Schools: Trends and Divergences; A Comparative Study on ECO-School Development Processes in 13 Countries*, edited by Finn Mogensen and Michela Mayer. Vienna: Austrian Federal Ministry of Education, Science and Culture.

OECD (Organisation for Economic Co-operation and Development). 2007. *PISA 2006 Science Competencies for Tomorrow's World: Executive Summary*. At www.pisa.oecd.org/dataoecd/15/13/39725224.pdf.

Oliveira, Alandeom W., and Troy D. Sadler. 2008. "Interactive Patterns and Conceptual Convergence During Student Collaborations in Science." *Journal of Research in Science Teaching* 45, no. 5: 634–58.

Patronis, Tasos, Despina Potari, and Vassiliki Spiliotopoulou. 1999. "Students' Argumentation in Decision-making on a Socio-scientific Issue: Implications for Teaching." *International Journal of Science Education* 21, no. 7: 745–54.

Ratcliffe, Mary. 1997. "Pupil Decision-making about Socio-scientific Issues within the Science Curriculum." *International Journal of Science Education* 19, no. 2: 167–82.

Sadler, Troy D., Sasha A. Barab, and Brianna Scott. 2007. "What Do Students Gain by Engaging in Socioscientific Inquiry?" *Research in Science Education* 37, no. 4: 371–91.

Sampson, Victor, and Douglas B. Clark. 2008. "Assessment of the Ways Students Generate Arguments in Science Education: Current Perspectives and Recommendations for Future Directions." *Science Education* 92, no. 3: 447–72.

Siegel, Harvey. 1988. *Educating Reason: Rationality, Critical Thinking, and Education.* New York: Routledge.

Simonneaux, Laurence. 2001. "Role-play or Debate to Promote Students' Argumentation and Justification on an Issue in Animal Transgenesis." *International Journal of Science Education* 23, no. 9: 903–27.

Toulmin, Stephen E. 1958. *The Uses of Argument.* Cambridge: Cambridge University Press.

UNESCO. 1977. *Intergovernmental Conference on Environmental Education: Final Report.* At www.gdrc.org/uem/ee/EE-Tbilisi_1977.pdf.

UNESCO INRULED (International Research and Training Centre for Rural Education). 2004–2010. "Education for Sustainable Development." At www.inruled.org/en/research/themes/a25999.html.

UNO. 2002. *UN Decade of Education for Sustainable Development 2005–2014.* At http://unesdoc.unesco.org/images/0014/001416/141629e.pdf.

Uskola, Araitz, Gurutze Maguregi, and María Pilar Jiménez-Aleixandre. 2010. "The Use of Criteria in Argumentation and the Construction of Environmental Concepts: A University Case Study." *International Journal of Science Education* 32, no. 17: 2311–33.

———. 2011. "Proceso de toma de decisión y dinámicas sociales de grupos de estudiantes universitarios en la discusión sobre un problema sociocientífico abierto." *Revista de Psicodidáctica* 16, no. 1: 123–44.

von Aufschnaiter, Claudia, Sibel Erduran, Jonathan Osborne, and Shirley Simon. 2008. "Arguing to Learn and Learning to Argue: Case Studies of How Students' Argumentation Relates to Their Scientific Knowledge." *Journal of Research in Science Teaching* 45, no. 1: 101–31.

Wu, Ying-Tien, and Chin-Chung Tsai. 2007. "High School Students' Informal Reasoning on a Socio-scientific Issue: Qualitative and Quantitative Analyses." *International Journal of Science Education* 29, no. 9: 1163–87.

Yang, Fang-Ying. 2004. "Exploring High School Students' Use of Theory and Evidence in an Everyday Context: The Role of Scientific Thinking in Environmental Science Decision-making." *International Journal of Science Education* 26, no. 11: 1345–64.

Zohar, Anat, and Flora Nemet. 2002. "Fostering Students' Knowledge and Argumentation Skills through Dilemmas in Human Genetics." *Journal of Research in Science Teaching* 39, no. 1: 35–62.

12

Social Values and Sustainable Practices among Basque Inshore Fishermen

Pio Perez Aldasoro

The University of the Basque Country (UPV/EHU)

This chapter examines the ways in which the production practices and values of inshore Basque fishermen converge with a widespread interest in the care and protection of the marine environment. It investigates the correlation between social values and the environment. Or rather, this work analyzes how fishermen resolve the apparent contradiction between maintaining the productive activity of fishing, an activity that depletes resources, while still defending its future for coming generations.[1]

While a detailed analysis of practical knowledge in various Basque fishing ports might bring to light the dissimilarities between them, seen together the framework of sustainable practices in Basque fishing communities remains very similar at a fundamental level. This paper seeks to highlight only the main aspects concerning the development of sustainable practices; its analysis pertaining to the multiple aspects of fishing activity remains relatively cursory. To this end, a small number of the characteristics and patterns of inshore fishing on the Basque coast are condensed. Equally, this paper considers some important milestones—key characteristics in Basque inshore fishing—and then links these to the hypotheses set out in this work.

Fishing and Environmental Issues

Ecological Niche, Production, and Values

The characteristics that largely determine an ecological niche are the use and exploitation endeavored by humans. Thus, the relationship between an

1. The reflections that appear here are from the doctoral fieldwork of Perez Aldasoro in 2000, as well as subsequent postdoctoral research on attitudes and values in the fishing port of Hondarribia (Gipuzkoa).

ecological niche and humans tends to be characterized by reciprocity (Biersack 1999, 9). That is, an ecological niche—or the space where human activities evolve—is largely anthropogenic, having been transformed throughout history by human communities. Moreover, there exists a set of constraints operating between the environment and human inhabitants (Sanmartin Arce 1982, 46).

Humans receive the majority of their needs for survival from such an ecological niche, but the environment is also where they live, constituting their natural world. This terrain is echoed within the development of human social relations, as well as in the ways individuals come to embody certain desires and frustrations. Humans maintain an *aesthetically affective* relationship with their respective ecological niches, and this is the space that manifests the unique aspects of a cultural system (Tolosana Lison 1973).

The different techniques that fishermen have used to carry out their work throughout history are the result of putting into practice a body of accumulated knowledge, passed from generation to generation among fishermen, as well as the social values developed around the ecological niche. One of the most important processes of mediation between a social group and the environment is production, defined as the means through which people are organized to appropriate and transform an aspect of the natural world. Production, moreover, involves human labor to create goods that meet individual and collective economic needs. The production process (i.e., the conjunction of organic production, circulation, and consumption) involves certain forms of labor, capital, and natural resources that are determined according to the product to be obtained (Castro and Lessa 1982). Thus, all production involves a kind of degradation, which is expressed in resource depletion and pollution, among other outcomes.

Within the interaction of social processes and nature, it is possible to distinguish two elements that unite each dynamic: (1) the forms of appropriation and transformation of nature through work, and (2) the technical strategies used for such appropriation. In effect, technique is the instrumental link between nature and society. Given that it is inextricably linked to work, technique refers to the various forms and methods used upon the natural environment. In a similar vein, Talcott Parsons argues that technology is "the socially organized capacity for actively controlling and altering objects of the physical environment in the interest of some human want or need" (Parsons 1966, 15).

Environmental degradation is often the direct or indirect consequence of the application of inappropriate technology; namely in this case, some inappropriate fishing techniques that only take into account the criteria of efficiency and economic profit. These activities fail to appreciate the natural rhythm and functioning of the ecosystem upon which they are applied. While technologies can help meet demands, they can also kill, poison, and strip resources.

The combination of work, technical, and natural resources makes up a complex system within which a variation of one of its elements necessarily affects others. The definition of an environmental problem, in a particular space and time, is thus linked to the core defining features of social organization in its interaction with the given environment. Therefore, fishing, in any instance, cannot be seen as an activity removed from historical and social reality. Moreover, it is inextricably linked with every community and area around the Basque coast.

Ownership of the Sea and the Restriction of Access to Resources

In nonfishing communities, there existed a traditional concept avowing that the ocean was a collective resource and, therefore, access to it was free, away from the canons of private property. This concept dovetailed with the idea that the marine environment was an inexhaustible source of wealth. These two views are not shared by fishing communities, in as much the communal character of the marine environment has developed as a series of mechanisms that regulate free access to marine resources (Lopez Losa 1996). These mechanisms gain their significance in producers' organizations that regulate activity and are governed by the producers themselves. One model of this style of organization is that of the fishermen's guilds (*cofradías de pescadores*).[2]

Although the ocean is theoretically free and the assumption is that fishing is comprised of open access to exploitation, the realities of practice diverge from this somewhat. In fact, fishermen develop mechanisms to restrict the access of other competitors through various methods. Thus, in practice, fishing is not a private resource, but neither is it an entirely free resource. Fishermen do not own the ocean, but through the practice of their work, they may acquire certain rights of use (Corn Alkorta 1993).

Basque Inshore Fishing

It is virtually impossible to date the beginning of fishing as an economic activity in the Basque coast. There are, however, archaeological remains that prove its existence in the pre-historic era (Carcamo Grace 1996). There is no

2. Essentially, *cofradías* or guilds of mariners emerged in the Basque Country at two distinct moments in history (Erkoreka 1991): when the need first arose to address a growing number of challenges and problems related to fishing. Shortly thereafter, Cantabrian fishermen began to cluster into collective guilds, known as *cofradías* and the oldest fishermen's guild in the Cantabrian coast originated in the Middle Ages. Then, in the thirteenth and fourteenth centuries nautical guilds were established in Hondarribia (Fuentarrabía), Donostia-San Sebastián, Deba, Lekeitio, Bermeo, and Plentzia.

doubt that fishing is one of the region's oldest and most productive activities and may well explain the entire history of human settlement patterns in the coastal marine environment (Rosique and Rebato 1997).

Although fishing is currently enduring tough times (Caja Laboral Popular 2010), the Basque people maintain a deep connection to its culture. Symbolically, fishing has created ethnic and cultural identities; this is reflected in celebrations, rituals, and in the collective imagination (Caro Baroja 1978).[3]

Until the mid-twentieth century, inshore fishing was unsystematic, deploying only a very low level of technical applications. For example, the reflection produced by the movement of fish (known as *ardora*) and the presence of predators (known as *manjua*) helped fishermen to detect the presence of their catches. It also showed them where to cast nets and rods so that their efforts were fruitful (Arbex 1978). Nowadays, technological advances in fishing have facilitated the detection and location of fishing grounds.

At present, the inshore fleet consists of 205 vessels representing 76.6 percent of the total Basque fishing fleet and 0.27 percent of all fishing vessels of the European Union (EU).[4] In addition, inshore fishing employs the most people in the sector as a whole, or a total of 1,484 jobs (56.6 percent of all Basque fishermen). The inshore fleet, however, has smaller and less powerful boats. The sector is experiencing serious difficulties for survival given large price variations, quotas set by the EU, and the need to intensify efforts in order to cover fixed costs (Caja Laboral Popular 2010). The 205 vessels form two distinct groups—the traditional fleet and the surface fleet—that differ based on vessel size, the type of fishing gear employed, and species caught.

The Traditional Fleet

The traditional fleet is composed of about seventy-three vessels, with a gross register tonnage (GRT) of under fifty register tons and crews of up to five people. The most common fishing method is long-lining (*palange*), in which several fishhooks are suspended from one line. The most important catches are hake (*Merluccius merluccius*), sea bream (*Archosargus rhomboidalis*), and mackerel (*Scomber scombrus*). Small boats carry out bottom fishing with small crews. These types of boats perform their tasks both day and night. Many types of fishing gear are engaged, including pots, lines, rods, and small nets, among others.

These boats start their season in late February. The first months are devoted to fishing for hake, horse mackerel (*Trachurus trachurus*), and mackerel—all with a hooked fishing rod (*al anzuelo*) until mid-June. Later, some of

3. A recent study of the Basque economy shows that fisheries have reduced activity by about 50 percent in the last twenty years (Caja Laboral Popular 2010).

4. Data provided by the Department of the Environment, Planning, Agriculture and Fisheries of the Basque Government in September 2010.

the vessels engage in fishing tuna (*Thunnus alalunga*) with a trolling system. They practice this fishery until mid-October. At the end of the season, some vessels may engage in collecting subsidiary algae.

The Surface Fleet

The surface fleet is composed of around 132 vessels, weighing over fifty register tons with an average crew of twelve people per boat. Vessels engaged in surface fishing work two seasons that form the axis of its activity: fishing tuna and inshore anchovy. While this fleet is dedicated to fishing anchovy (*Engraulis encrasicolus*) and tuna, it can also render significant catches of mackerel, horse mackerel, and sardine (*Sardina pilchardus*). The fleet's season is between March and October (March to June is anchovy season; July to October is tuna season). Anchovy fishing boats leave daily from the harbor at sunset, fishing at night and returning to port in order to sell their catch in the early hours of the day. The gear used is purse seines netting. Once the tuna fishing boats depart from port, they stay at sea for a couple weeks. The number of days at sea will depend on the supply of bait and the amount of tuna caught. Tuna is caught individually using baited rods.

Gear and Sustainable Fishing Techniques

There are many advantages offered by sustainable fishing gear. The main benefit of this kind of fishery is related to the specificity of the actual catch, since no harm is done to any other species. Moreover, no smaller fish are captured; thus, the discarding of fish (that could be used later as bait) overboard is almost nonexistent. By using these fishing techniques, the seabed does not suffer any alteration, nor do they damage the marine ecosystem.

Long-line Fishing

A long-line consists essentially of a main line that carries many hooks attached to it by other lines (*brazoladas*). The hooks are spaced apart at a given distance depending on the length of the ganging (so that they do not cross) and depending on the species to be captured.

There are long-lines of various sizes used in fishing, from those measuring only 100 meters, to those in excess of 60,000 meters. In the traditional inshore Basque fleet, a long-line rarely exceeds 1,000 meters. Long-lines are also composed of accessories needed for its floating, anchoring, and signaling. The long-line is cast above a given height of the ground that depends on the distribution of these accessory elements. Long-lines for bottom and surface fishing depend on the species to be caught and the season.

In order to rig the long-lines, when cast, the vessel is maintained in the proper course and at a speed of up to 9 knots, according to sea conditions. At the first pass, a buoy is hoisted to a flotation device that is tacked to the lead winch that then passes through the drum. The rigging operation continues until the collection of the end of the line head.

Fishing by Purse Seine Netting

The purse seine netting gear is used mainly in Cantabrian anchovy fishing, but also to catch lesser fish, such as horse mackerel, sardines, or mackerel. As its name suggests, the "purse" seine net is designed to encircle a school of fish by forcing them to stay within the circle formed by a large net. Once the pod is inside the circle, the net narrows, capturing the fish in a pouch.

Once the school is located, the vessels proceeds in order to determine the starting point for rigging the net. To this end, it is necessary to know precisely the size, speed, and direction of travel of the school; prevailing wind; and current; among others. (The size of the gear and maneuverability of the vessel are known in advance by the skipper.) To calculate the first three elements, the boat crosses the school several times, always coming from behind. In each pass, it leaves a buoy in the center of the school in order to take bearings that serve to confirm its direction and speed. This operation can also verify the density of fish school.

Once placed in the starting position, a dinghy sets out and grabs the fore fist of the net with a corresponding throw starboard. The seiner navigates in a circle and the gear is cast while simultaneously arriving at the starting position. Aided by a hydraulic pulley casting, the dinghy collects the net right after it is cast and begins to tack it in order to close the circle. The dinghy moves along the side of the "cod end" and incorporates part of the headline. Once the cod end is closed, dip nets or large vacuum tubes are used to load the fish on board.

The nets that Basque fishermen use are rectangular, having a length of approximately 300 meters and a height of 70 meters. With regard to technical measurements of the nets, one must take into account the species to be captured and the size and power of the given vessel, as these technical differences influence the choice of the dimensions of the rigging.

Tuna Fishing Using Live Bait

Fishing with live bait was first used by Basque fishermen in Spain in 1950, after its effectiveness was realized in the neighboring ports of Hendaia (Hendaye) and Donibane Lohizune (Saint Jean de Luz) in the French Basque Country. In turn, this fishing technique was also exported to the coast of California (Ingelmo 1984). Tuna is caught one by one, using small, live sardines, anchovies, or horse mackerels. These baits are kept alive on board in nurseries that are composed of huge tanks and that maintain a constant flow of seawater.

First, fishermen must locate a school of tuna. Once found, a few live bait-fish are tossed into the water so that the tuna approach the boat. Meanwhile, the other anglers bring out a thin hose that sprinkles "rain" over the water and prevents the tuna from seeing the boat. Then, using a bamboo cane with a short line, the fishermen release a few live fish that have been hooked. Next, the tuna (a very voracious fish) go into a feeding frenzy biting anything that shines or moves, ultimately ending up caught on the hooks.

The school then piles up alongside the boat and the fishermen then catch them one by one. The duration of the fishing depends on the tuna themselves. Once their appetite is satiated, or they have been scared for any reason, they cease to bite the hooks and move away. This fishing requires great concentration and coordination from the crew, because if the tuna scare or escape the hook, it is likely to be caught by others or to seek shelter deep at sea.

In addition to this active fishing technique, there is a variation exploited by some boats in Hondarribia called *trolling*, or the troll (Burrutxaga 1992). The technique involves dragging bait or a lure from the boat so that the fish are tempted. The boat speed adapts to the dynamic characteristics of the fish to be captured. In the case of tuna, a fish that is a good swimmer, the boat can cruise from 5 to 8 knots. This fishing method is most often used when fishing for bighorn or Atlantic bluefin tuna (*Thunnus thynnus*).

Unsustainable and Over-exploitative Fishing Gear

In 1980, the French introduced driftnets and pelagic trawls in the Bay of Biscay. In contrast to traditional methods, fishing with driftnets and pelagic trawls can be done throughout the year. Traditional techniques are necessarily limited to the time of year when the anchovy and tuna surface (i.e., spring and summer).

In trends starting with the use of driftnets and pelagic trawls, the catch from traditional fleets gradually declined, with decreases in several different periods linked to the technique that was used. Live bait catches began to decline after 1990, while long-line catches fell sharply between 1986 and 1991. The yield of bait catches plummeted in 1991 and 1992 (Macias and Ferreiro Ares 1997). The lowest catch was recorded in 1999.[5]

Various target species, including hake, disappeared from traditional catches with the arrival of the pelagic trawls. With the disappearance of hake, only two economically viable fish remained for traditional fisherman in the Bay of Biscay: anchovy and tuna. Both species showed signs of attenuation, due to declining catches. Furthermore, these unsustainable techniques had a

5. For telling reports on fleet catches during these years, see http://www.azti.es/en/responsible-fisheries-management.html Basque.

negative impact on some nontarget species, such as dolphins and other cetaceans, which were often found floating dead in the sea (European Commission 2004, 10).

Driftnets

Driftnets or drift fences (*volantas*) are made of polyamide multifilament and have a density appreciably higher than the seawater. They are composed of successive chain links up to 100 by 100 meters in measurement. Thus they can be adjusted to reach a variety of depths. Once placed in water, the nets remain floating on the surface by a headline, boltrope, or a line of floats so that the net falls and hangs down toward the bottom of the water. It remains upright weighed down by a second headline measuring about 20 feet below the water.

Lighted buoys are attached to the ends of the nets, since they operate at night in areas between 3,000 and 5,000 meters deep. At the outset, given the location of the nets and their tendency to drift unchecked upon the ocean's surface, their use constituted a significant obstacle to normal navigation and more likely than not to boats and ships in general. Further, there has been a direct drop in the catches of Basque inshore fisherman that has apparently been caused by the navigational interference from driftnets, as the lower yields have been paralleled by increases in catches from the driftnets.[6]

The main difference with the use of traditional gear versus the use of driftnet is that the latter is an entirely nonselective method, especially with respect to the species of fish caught. This inevitably leads to a significant capture of diverse marine species, among which a remarkable number are eventually discarded and usually dead.

The driftnet strategy is based on a widespread exploitation of a great mass of the water column. This differs greatly from the selective technique used by the traditional fleet, which yields a higher quality capture of fish, both in appearance and freshness. That is, the fish caught using driftnets usually stay dead for several days under water, consequently degrading the quality of their meat.

The Pelagic Trawl

The trawls belong to what is considered mobile gear. Whether it is for bottom or surface use, the trawl is towed by one or two boats that steer into schools of fish, penetrating their core, so that the fish get caught in the cod end of the trawl. Vessels using pelagic trawls look for large, dense, and less active

6. This argument is based on data in the work, *El Libro Blanco de la Pesca*, a study conducted and published by the Spanish Ministry of Agriculture, Fisheries, and Food in 2008.

schools of fish, given that pelagic fish are good swimmers and have highly developed visual and auditory senses. This gives them the ability to recognize and avoid the trap more easily, especially during the day and near surface layers of water.

Trawl nets are composed of four or more flat planes. The most common ways to use these nets are by shaping the planes into rectangular, oval, or circular form. Whatever form employed, the mouth is left so that it opens very wide. Its wings are short, and its peak (unlike benthic trawls) is located on the bottom to prevent the escape of fish from below.

Some Notable Milestones in Basque Sustainable Fishing

The War of the Anchovy and the War on Driftnets

In the early 1990s, there were various confrontations in the Bay of Biscay—known as the "war of the anchovy"—fought between inshore Basque and French ships that were engaged in pelagic trawling. Other incidents occurred during tuna season, in which French ships using driftnets were confronted. According to Basque fishermen, uncontrolled development of high-capacity fishing gear with low or questionable sustainability records (such as driftnets and pelagic trawls) was contributing to a situation of overexploitation of certain species, such as hake and sea bream. Having depleted these species, the very same vessels were engaging in the exploitation of anchovy and tuna, the two pillars upon which the Basque coastal economy rested. The presence of these fishing techniques was substantially altering the yields of both species of fish caught with traditional sustainable gear.

In the war of the anchovy, in addition to rejecting the use of nonselective fishing gear, there was an unmistakable demand concerning the distributions of total admissible captures (TAC) and other fishing quotas. Under the treaty of accession of the Kingdom of Spain to the EU, it was believed that the Cantabrian fishing fleet employing traditional gear yielded 90 percent of the anchovy quota, while the French pelagic trawl fleet yielded only 10 percent. These percentages, however, were agreed between the French and Spanish based on criteria established outside the fishing world and influenced by convoluted political interests.

Basque fishermen argued that quotas were unsuitably applied to fishing fleets that had cared for their fishing grounds for centuries, while the government had not. They also maintained that the rights of access to fishing must go hand in hand with cultural practices and the economic needs of communities that have historically relied on its existence. According to the Basque fishermen, the use of responsible and selective gear should be a key element in establishing criteria for allocating quotas. At that time, in the early 1990s, political criteria took precedence over biological facts in establishing restrictive measures on catches. Moreover, these allocations not only failed to take into account traditional knowledge systems of fishing acquired by years of practice, but often ignored scientific knowledge concerning natural fishing resources.

Given low market prices, Basque fisherman refused to continue to enrich a fishing fleet that competed with them by using nonselective and harmful gear. Thus, they labeled Spanish and European governmental managers as responsible for the demise of anchovy during the 2006 and 2007 fishing season. In their opinion, the governments should have taken into account certain fishing methods and the impact of their implementation on a particular species.

Itsas Geroa

In 1995, groups of inshore fishermen from both sides of the Franco-Spanish border, and persons and entities concerned with the problems of the marine environment, came together to establish a basis for communication and exchange of information in order to work together to protect the future of the ocean. The results of this first meeting were public awareness campaigns against driftnets and the development of product labeling that let the consumer know how the fish product was caught (i.e., with safe and sustainable techniques).

The organization Itsas Geroa (meaning "the future of the sea" in the Basque language) is comprised of almost all the fishermen associations in Basque ports, as well as conservation groups and other Cantabrian and French advocates of traditional, selective, responsible, and sustainable inshore fishing. The regulatory statutes of Itsas Geroa explains that this partnership functions based on a common interest in defending maritime cultural and social values. Thus, it conceives of itself as an instrument that overcomes the conventional isolation resulting from the political, social, and geographical fragmentation of seafarers. Its principle aim is to contribute to the establishment of a social movement that offers answers and alternatives to the progressive global socio-ecological deterioration of the environment.

By creating a partnership orientated toward responsible fishing, fishermen drew upon centuries of struggle they had inherited from the Cantabrian Sea guilds (*cofradías*). The guild of mariners, as an institution, had long been aligned with objectives similar to professional labor unions. That is, its main objectives were aimed at regulating fishing activity through the control of labor, the fishing seasons, and the techniques employed (as well as the product marketing and sales). In this sense, the guilds function as institutions whose aim is to reduce, or to attempt to mitigate, the risks inherent in the fishing world. This was accomplished first by establishing measures to moderate the dangers connected to their professional activity, and second by creating their own supply and demand cycles by regulating fishing seasons, which creates the optimal conditions for profit.

The Charter of Cedeira

Another milestone concerning Basque fishermen in pursuit of sustainable fishing is their active participation in the drafting of the declaration called the Charter of Cedeira. The merit of the covenant was that it brought together a large number of geographically different guilds that had common problems: overfishing, risky technological improvements in fishing gear and vessels, poor selectivity of fishing gear, and the environmental impacts of the gear.

The charter's philosophy was based exclusively on the urgent need to take action and to recover the ocean. It stated that everyone should take responsibility, both the fishing sector and the administrative sector. Both stakeholder groups must be prepared to work together to reverse contemporary rapacious behavior by cultivating responsible attitudes toward the ocean and respecting its rich fishing grounds, marine ecosystem, and biodiversity. The charter also emphasized the need to develop sustainable business practices through a management model that, not only took into account the quantitative aspects of fishing resources, but also the need to maintain a healthy ecosystem as a prerequisite for recovery.

To achieve these proposed objectives, the stakeholders:

- Imposed an immediate ban on all surface and mid-water trawling
- Amended the ban to suit the areas closed to bottom trawling in the northwestern Bay of Biscay's fishing ground
- Reviewed the minimum sizes of target species
- Determined the maximum size of mesh nets
- Established fair work schedules and living standards for fishermen
- Began to monitor and control all vessels.

In order to curb overexploitation of fish stocks, "fishing strikes" and closures that scientists deemed necessary took place. Additionally, the stakeholders sought to upgrade vessels in order to provide the means to ensure that protective measures could be taken.

Conclusions

The ability and strength of fishermen is anchored in the traditional hold that fishing has on the Basque cultural world. The most significant aspect of this tradition is the large group of people devoted to working the ocean—be they sailors, shipowners, day laborers, or laymen. This human capital has been acquired over a great deal of time, along with a number of skills that allow inshore Basque fishermen to exploit the marine environment and its natural assets.

While no one in the fishing industry actively promotes nonsustainability, it is arguable that most fishing will eventually become unsustainable. Therefore, marine resources constitute an asset that must be preserved and protected. Each fishing season, every fisherman claims that his catch is less or that the number of fish is declining or that his profit is much lower than a decade ago. The fishing industry as a whole is steeped in a fashionable pessimism, and it is not easy to predict whether fishing itself will end before the fishermen do. In other words, fishing as a way of life must be preserved. Management strategies routinely give an excessive importance to the biological dimension of sustainability, while it would be advisable to pay more attention to other sustainability criterion.

The new and modern concept of sustainability must consider the ecological and socioeconomic dimensions linked to the communities that exploit natural resources. It is possible to discern, in the articulation between society and nature, two facets or forms of concern, one rendered by nature and the other by the social group:

1. The first refers to the specific material aspects of the link, the appropriation of elements from the natural environment, and the later processing and consumption of the elements captured. The different phases in the development of a society involve different forms of domination and control over nature. The degree and character of this control will depend on many factors, from environmental variables to economic and ideological concurrences.
2. In this latter manifestation, "the social" inscribes itself within the second facet. These are the conditions and characteristics needed to possess an *aesthetically affective* appraisal of the natural environment. Symbolic representations of nature acquire a unique aspect in each cultural system.

Both areas are linked in reality, permanently interacting and mingling with each other. As Alain Touraine (1982) argues, human action is, in sum, knowledge, labor, and judgment. The material world influences representations as much as the symbolic world. Therefore, emphasis should be placed on the material processes and appropriate material resources, without forgetting the cultural, ideological, and political influences that affect the process of articulation. In this sense, humans incorporate values and identities into processes of appropriation and transformation of nature from within the social realm. Nature is understood according to material and ideological forms, or according to particular conceptions that are generated in the evolution of the social group.

The fishing methods developed by Basque fishermen exploit the marine ecosystem of the Bay of Biscay while respecting its ecosystem. Due to their high selectivity, their catches are less numerous than those of industrial fishing, yet their catches have a higher unit value and a higher economic-profit ratio. Additionally, traditional fishing can be a positive factor in the conservation of biodiversity and marine ecosystems, given that the sustainable exploitation of inshore resources involves limiting other uses of the coast that have detrimental effects on ecosystems.

Basque fishermen are not a collection of isolated individuals; they have a history that binds them to a specific resource and a given ecosystem. They act collectively in order to organize access, exploitation, and protection of the resources they exploit. They have created the necessary institutions in order to self-regulate their catches. Thus, they have prevented not only the exploitation of resources and possible extinction of fishes, but have also enabled a collective, communitarian level of communication that may well ensure fishing as a way of life (Suárez de Vivero, Rodriguez Mateos, and Florido del Corral 2008).

Labor is the product of individuals moving within a social fabric that dictates norms and values. Coping mechanisms and behaviors are derived by the consistent guidelines set by the social group. The environment, in turn, will impose its conditions, allowing certain types of intervention over it. It is essentially simple elements that make up the labor process (work, object, and environment). At this juncture, it is possible to discern the social and human elements that will shape the various scenarios that build on the relationship between society and environment.

In this sense, it is important to note the survival of the guild as an institution. Despite their problems and difficulties, they represent communities that have historically been linked to the ocean and that refuse to allow an uncontrolled exploitation of marine resources. They also form a group clearly defined through a close interaction between economic activity and ecological fishing specific to their social and cultural organization. However, the rejection of the modern techniques of fishing—whether trawling or driftnets—is not a rejection of modernity; rather, it is a commitment to the future.

The mere fact that certain groups of fishermen (through their organizations and without the participation of state institutions) are seeking solutions to their problems should be considered a major step forward. This demonstrates that there is still a rationale for the survival of these groups in the changing contexts of the fishing world. It should also be noted that the groups' influence on fishing has affected the whole of the community, as well as made manifest a sense of strong awareness among neighboring communities. It is virtually impossible to find fish in the local markets that have been captured by overexploitative gear. Equally, the practice of selling fish below set size and weight limits has virtually disappeared from fishmongers' shops.

References

Arbex, Juan Carlos. 1978. *Arrantzaria: Evolución de la construcción artesanal de la pesca y de sus artes en Euskalerria*. Bilbo: Petronor.

Biersack, Aletta. 1999. "Introduction: From the 'New Ecology' to the New Ecologies." *American Anthropologist* 101, no. 1: 5–18.

Caja Laboral Popular. 2010. *Informe sobre economía vasca 2009*. Mondragón: Servicio de publicaciones de Caja Laboral Popular.

Caro Baroja, Julio. 1978. "Los vascos y el mar." In José Miguel Barandiaran, *Itxaskaria: Los vascos y el mar*. Bilbao: Petronor.

Castro, Antonio Barros de, and Carlos Francisco Lessa. 1982. *Introducción a la economía: Un enfoque estructuralista*. Translated by Claudio Colombani. Madrid: Siglo XXI. Portuguese version: *Introdução à economia: Uma abordagem estruturalista*. Rio de Janeiro: Co. Forense de Artes Gráf., 1967.

Erkoreka, Josu Iñaki. 1991. *Análisis histórico-institucional de las cofradías de mareantes del País Vasco*. Vitoria-Gasteiz: Eusko Jaurlaritza.

European Commission. 2004. *European Code of Sustainable and Responsible Fisheries Practices*. Luxembourg: Office for Official Publications of the European Communities.

Gracia Cárcamo, Juan Antonio. 1996. "El sector pesquero en la historia del País Vasco: Esbozo de los conocimientos actuales y problemas abiertos a la futura investigación historiográfica." *Itsas Memoria-Revista de Estudios Marítimos del País Vasco* 1: 209–14.

Ingelmo, Alfredo. 1984. *Inventario de artes de pesca en Euskadi*. Vitoria-Gasteiz: Gobierno Vasco.

Lisón Tolosana, Carmelo. 1973. *Ensayos de antropología social*. Madrid: Editorial Ayuso.

López Losa, Ernesto. 1996. "Derechos de pesca, privilegio y cofradías de mareantes: Una aproximación histórico-institucional a la gestión de los recursos pesqueros en el País Vasco (XIV-XIX)." In *Cambio Institucional e Historia Económica*, edited by Josep Pujol, Pedro Fatjó, and Neus Escandell. Barcelona: Universitat Autònoma.

Macías, Luis Miguel, and Carmen Ferreiro Ares. 1997. "El sector pesquero vasco: Situación, evolución y retos de futuro." *Papeles de Economía Española* 71: 62–85.

Maiz Alkortza, J. Agustín. 1993. *El sector pesquero vizcaíno 1800–1960: Análisis de la interacción de los elementos ambiental, extractivo y comercial en la pesquería*. Vitoria-Gasteiz: Departamento de Agricultura y Pesca, Gobierno Vasco.

Ministerio de Agricultura, Pesca y Alimentación. 2008. *El Libro Blanco de la Pesca*. Madrid: Servicio de Publicaciones del Ministerio de Agricultura, Pesca y Alimentación.

Parsons, Talcott. 1966. *Societies: Evolutionary and Comparative Perspectives*. Englewood Cliffs, NJ: Prentice-Hall.

Pérez Aldasoro, Pío. 2000. "Pasaiko arantzaleen lan eta gizarte harremanei buruzko azterketa antropologikoa." PhD diss., University of the Basque Country (UPV/EHU).

Rosique, Javier, and Esther Rebato. 1997. "La adaptabilidad humana en las comunidades pesqueras." *Zainak. Cuadernos de antropología-etnografía* 15: 265–77.

Sanmartín Arce, Ricardo. 1982. *La albufera y sus hombres: Un estudio de antropología social en Valencia*. Madrid: Akal.

Touraine, Alain. 1982. *El postsocialismo*. Translated by Janine Muls de Liarás. Barcelona: Planeta. French version: *L'Après socialisme*. Paris: Bernard Grasset, 1980.

Santiago Burrutxaga, Josu. 1992. *La pesca del atún blanco en Euskadi: Campañas 1990–1991*. Vitoria-Gasteiz: Gobierno Vasco.

Suárez de Vivero, Juan Luis, Juan Carlos Rodríguez Mateos, and David Florido del Corral. 2008. "La gobernanza en la pesca: De lo ecológico a lo ético, de lo local a lo global." *Scripta Nova* 12, no. 278 (online journal). At www.ub.es/geocrit/sn/sn-278.htm.

13

Sustainable Development and the Values of Well-being and Globalization

EDUARDO RUBIO ARDANAZ
University of the Basque Country (UPV/EHU)
JUAN ANTONIO RUBIO-ARDANAZ
University of Extremadura
XIAO FANG
Lu Xun Center for Chinese Studies

Sustainable development is a concept that covers everything from well-being to globalization in any discipline related to the study of economic development. The concept encompasses questions concerning international effects and implications in the present and in the coming decades, and helps us to explain what is happening at a local, regional, and international level.

The terms *globalization* and *sustainable development* refer to inevitable phenomena. Globalization is an unstoppable process whose effects have been felt for decades and whose pace has accelerated over the years. Sustainable development emerges in response to the need to preserve and improve the continuity of our societies and cultures for the inhabitants as well as for future generations.

It is clear that sustainable development has increasingly gained in scope based on the idea of the future: without it there can be no future for humans on this planet. In this sense, there is also another aspect that is equally inevitable and centers around a framework of consensus and participation of all social, cultural, economic, and political actors. By contrast, globalization seems to be characterized by inexorability (given that it presents itself to us from the past at a growing pace), and by the imposition of a process in which most socioeconomic, cultural, and political actors do not participate in decisions, even though they suffer their consequences.

The idea of well-being also appears within this framework and is a concept that is applied when making decisions at a local–municipal level and represents the overriding objective of sustainable development itself (City of Winnipeg 1995; Cuthill 2002; Blanchflower and Oswald 2004; Blanchflower

2005). Participation, in this sense, emerges as an objective of sustainable development as well as has implications for and connections with well-being. We consider these questions unavoidable when facing the true challenges of the future in relation to the concepts presented.

Our reflection is based on the basic ideas of globalization and sustainable development and captures their most significant characteristics, thus helping us to measure their consequences, objectives, and evolution. We will also carry out an analysis of the main critiques of these phenomena. We would also like to consider the ways in which sustainable development is materializing from a global perspective and in relation to changes in individual and social values.

A Review of the Concepts of Globalization, Sustainable Development, and Well-being

Globalization is not a recent phenomenon. We can trace its advance through the centuries and note its continuous accentuation. Insofar as we understand the question as it stands at present, we can distinguish two phases of globalization: that which starts at the end of World War II in 1945 and that which corresponds to the oil crisis of 1973. The second phase, which endures today, has been more acute, with a pace of growth that is perceived as unstoppable and out of control. During this period of time, the great influence produced by recent technological revolutions, information technologies, and communications is clearly evident. It reaches its dialectic and conceptual peak in the 1990s (Keeling 2004).

The first signs of globalization are found within the financial sphere and extend to the fields of information and the economy, in general. Also, it would be impossible today to separate the phenomenon of globalization from its environmental, political, social, and cultural effects (CEPAL 2002; Lucas 2004). In all of these areas we find four characteristic features that we will go on to analyze here in virtue of their relationship to sustainable development:

- *Globalization is irreversible.* Due to the level of technological development attained at this point of the twenty-first century, it would be impossible to return to levels of isolation similar to those of past decades, especially with regards to transportation and communications. This does not mean that globalization should be assumed and, above all, accepted in a noncritical way. Hence, globalization, considered as irreversible, must be the object of careful consideration in terms of its processes and consequences.
- *Contradictions exist within globalization.* We should understand globalization as a neutral phenomenon that has both positive and negative overtones, depending on how it is managed and directed

(Lucas 2004). Therefore, in and of itself, globalization is not a negative phenomenon (although many of its consequences are, especially due to the neoliberal economy in which it evolves). Neither is it the best of all possible realities (even though some of its consequences have been positive). It is quite likely that no single form of globality exists and that we must aim for the most convenient and respectful form for all of the inhabitants of the planet (Albert 1993; Rodrik 2001).

- *Corresponding fields of influence are at play in globalization.* International influences have repercussions on the regional and local levels. And, vice versa, the influences of the local and regional levels can be felt at the international level by way of mutual repercussions. Fields that were previously unconnected now receive clear influences from each other. This increasingly occurs at a greater speed, inexorably and with a certain lack of control.
- *Inequality exists among those who participate in globalization.* Globalization leads to countless imbalances that many inhabitants of the planet suffer from today. The main agents (governments and multinational companies in particular) impose their interests on the less powerful and poorly organized agents. Only that which is capable of influencing the economic results of the main agents acquires the status of reality, while that which is outside the interests of the main sociopolitical and, above all, economic agents is undermined, ignored, and even rejected as irrelevant.

Within this framework, we would like to introduce the idea of well-being, a controversial concept that is defined in many different ways. In fact, there are as many definitions of the phenomenon of well-being as there are instruments to measure it. We begin from the premise that well-being has subjective (self-assessed) and objective (ascribed) dimensions. It can be measured at the levels of individuals or society. And to a great extent it accounts for elements of life satisfaction that cannot be defined, explained, or primarily influenced by economic growth (McAllister 2005, 2).

Critiques of Sustainable Development

A considerable number of critiques have been put forward with regards to the concept of sustainable development over recent years and since the appearance of the Brundtland Report (World Commission on Environment and Development 1987). We will now outline the main arguments and approaches that have been critical of sustainable development and its implementation, and that have gone as far as to suggest that the concept of sustainable development is nothing more than a utopian vision rather than a map for making practical decisions; with the consequences that such an idealistic vision implies.

Against the Multiplication of Economic Development

The first of the critiques focuses on the demand made by the Brundtland Report that development be extended to the whole of humanity (by between 5 to 10 times, as we saw previously). If the development that existed in 1987 already constituted a serious problem for environmental sustainability, the idea of increasing it by the proportions recommended would represent a threat that would be disproportionately greater. This critique is amplified if we consider that the human population is growing at a vertiginous rate, which means that the needs of a growing population are also growing, even though they may remain stagnant at an individual level. Therefore, in order to reach the entire human population of the present day, the provisions of the Brundtland Report would already have fallen short (despite already being of great magnitude).

In response to this criticism, we will present two different solutions. On the one hand, we could trust that technical advances will enable the efficiency of production methods to be multiplied in such a way that the impact of more economic development will result in the same environmental impact as now, or even less. On the other hand, we could believe that we can somehow find a new way of developing and organizing ourselves that would imply less of a burden on the environment and less resources for the planet that would enable development to be extended to the whole population of the world. This critique should not imply the limitation of development to designated areas of the planet, leaving the rest underdeveloped with the excuse of limiting the expansion of development. Neither should we forget that one of the solutions proposed for this situation would inevitably imply a drop in the quality of life of those who consume too much, so that the rest of the human population would then be able to attain decent levels of development.

The Danger of Associating Sustainable Development with Economic Development

The second critique is based on the speed at which governmental and company ranks associate "development" with "economic development." By doing so, they can thus justify their commitment to sustainable development without taking into account the other constitutive elements of it. In fact, these (social and environmental) elements constitute the hard core of the Brundtland Report, which focuses on people and the civil and social rights that they are entitled to.

Thus, what is of utmost importance for the Rio Declaration (1992) is the human being, who is recognized as having the right to a healthy, productive life in harmony with nature. In any case, although this objection is not insurmountable or irremediable, it does represent a wake up call of the first degree to ensure that we do not become distracted from the true meaning expressed in the Brundtland Report. Above all, sustainable development should be underpinned by social and environmental concerns as expressed in the report.

Ambiguity of the Concept of Sustainable Development

In relation to the previous critiques, we can also present a third that attributes a certain ambiguity to the conceptualization and definition of sustainable development in such a way that it ends up depriving it of all meaning, rendering the term applicable to any kind of act or process, whether it be truly sustainable or not. The risk that activities contrary to the spirit of sustainable development end up being labeled as such is a reality and a mere game of language and of business or governmental marketing that seriously jeopardizes the viability of the process (Robinson 2003, 844).

This ambiguity could lead to people or groups simply looking out for their own interests and justifying them by emphasizing the part of the definition that best suits these interests: environmentalists want environmental systems sustained, consumers want consumption sustained, and workers want jobs sustained. As Richard B. Norgaard (1988, 607) notes, "with the term meaning something different to everyone, the quest for sustainable development is off to a cacophonous start."

In response to this criticism, some authors, recognizing the breadth and vagueness of the concept of sustainability, see this more as a positive aspect of political opportunity than a problem in and of itself. The scope of the definition enables dialogue and a potential meeting ground to be established for positions that are initially opposed and that would no doubt be less possible if a more exact definition was available (Robinson 2003, 844, 849–52).

Opportunism or Sustainable Development

In light of this, a large majority of those who discuss sustainable development (at a governmental and business level), brandishing the term within their speeches and practices, do so in ways that are not entirely honest in order to gain the favor of a consumer public that is increasingly sensitized to the situation. It is even used by companies in an "attempt" to satisfy the "ecological" demands of consumers, thus managing to attain even greater levels of consumption (Morton 2000). In other words, this involves a tactical posture as opposed to a clear will to promote sustainability, behind which economic interests are carefully hidden (Robinson 2003, 842).

The proliferation of standards and certificates of sustainability (based on environmental friendliness and social responsibility) may be motivated by this commercial intention. In fact, such standards are being applied to a growing number of products and services that have emerged over recent years. On the other hand, the attraction of sustainable development for a great majority of governmental and business agents resides precisely in the "business opportunity" that their application represents.

Sustainable Development as a Screen that Hides the Real Problems of Humanity

The last critique, and the most serious of all, focuses on the undesirable effect that the dynamic of sustainable development could have on the real problems of the world today: we may stop appreciating real needs by being distracted by other problems. Sustainable development would thus function to introduce successive changes while "threats to the way power is currently distributed and held" are contained by those in power (Lohmann 1990, 82). By ignoring what is most important, sustainable development would, therefore, be reformist by not questioning either the distribution of power or situations of exploitation, and by ignoring questions such as the need for a political and social change (the antiglobalization movement can be located on this side of the critique) (Klein 2000).

Those who hold this hard-line view cite indicators (provided by international organizations) that clearly point to a widening of the gap between the rich and the poor across the world, both at an individual and international level, despite the increase of economic development and the implementation of actions aimed at sustainable development (Fernando 2003, 9). This accusation is extremely serious and certainly the most difficult to tackle among all of the critiques presented here. In any event, these different critiques do not detract us in any way from a reality that demands a response. It is also important to bear in mind the specific applications in which sustainable development may ultimately be put to effect.

The Relationship between Well-being and Participation

The need for participation extends not only to all areas of sustainable development, but also to the concept of well-being and the applications of it that have been carried out over recent years. As we see in the process carried out in the community of the City of Winnipeg (Manitoba, Canada) and in processes carried out in other places, the procedural cornerstone of them all is the participation of citizens and all of the representative actors of the community.

All of these agents and individual citizens are required to bring along their points of view and their legitimate interests in order to define and establish a framework for the objectives and standards to be set in an attempt to achieve these goals. Therefore, what all of these processes have in common is the will to continue and follow through (participation). In order to be able to act on an equal footing, all participants must have the information needed to be able to consider and make decisions that will reflect their legitimate interests, true values, and different points of view.

It is also worth noting that a correlation can be seen between greater indices of satisfaction and well-being in societies that are governed by democratic systems (as we understand these in the West) than in other countries; in those societies participation (although far from being perfect) is comparatively high.

In the different reports that we have examined (e.g., the City of Winnipeg [1995]), the subjects of participation and governance are situated at the same level as achieving greater quality of life and the sustainability of a new project. In fact, we see that participation and governance are among the points that are worked on the most and that are most effectively developed, with concern for them corresponding to the provision of budget issues and the procurement of technical means as well as the technical experts capable of developing them.

The concept of well-being is controversial and defined in many different ways (McAllister 2005, 2, 59). The same occurs with the concept of participation and especially with respect to its application (Harrison, Munton, and Collins 2004, 904, 906–7). Both concepts are closely interwoven in their application to sustainable development in different places across the globe. Thus, the application of the former necessitates, or is at least implemented through, the use of the latter (achieving well-being through participation).

The fact that programs and indicators of well-being are being drawn up in a way that is as participative as possible is helping to identify and make the objectives marked out by the concept of sustainable development more specific and measurable (Keyes 1998, 133; McAllister 2005, 14). Well-being also often implies other measures of change, whether political or relating to behavior and attitudes, which would be difficult to achieve without the active participation of citizens (Campbell and Jovchelovitch 2000; Doelle and Sinclair 2006). This participation is directed toward decision-making processes and not only toward the behavior to be corrected. As some authors clearly indicate, this is the only way to involve citizens in a problem that tends to be seen as one that is distant from them and that tends to be left in the hands of politicians and the authorities (Harrison, Munton, and Collins 2004, 914–15).

The Need to Continue Reflecting on Sustainable Development and Globalization

We, therefore, see that sustainable development and globalization are concepts that, paradoxically, are as incompatible and antagonistic to each other as they are inseparable and necessary. While globalization acts in a top-down way (the global influences the regional and local; the regional influences the local), sustainable development may act from the top down as well as from the bottom up (the local influences the global and the regional; the regional influences the global). In order for sustainable development and globalization not to be destabilizing forces in the places where they are implemented, it is important that the influences of globalization reach the different destinations in equal conditions so that they can be chosen or rejected and so that similar results can be obtained from the influences in the different places in question.

If this is not the case, the result will be the degradation of the weakest cultures as well their further lagging behind economically and technically; the acculturation of those who do not control the flow of information or the economic mainstream and its values. It is, therefore, also important that the resulting flow can be equally developed and applied subsequently, thereby avoiding the creation of frustrations among those who, having the capacity to develop, are rendered incapable of exercising their capacity due to unfair game rules at both an international and local level.

In the interests of sustainable development, globalization should be restrained and limited in cultural, social, and legal terms, especially in situations in which it falls outside the two premises mentioned previously for its application. If not, the consequences will be in total opposition to what sustainable development signifies and to its demand for participation and real decision-making mechanisms.

Finally, sustainable development should not signal the continuity by other means of a system that has been shown to be unfair and lacking solidarity, as well as harmful to the environment and even unsustainable. Rather, it should give way to a different world that is not only capable of being sustained in the future, but that also deserves to be.

The Application of Sustainable Development: Values versus Technical Efficiency

The varying ways in which different authors have proposed that the objectives of sustainability outlined previously be achieved are related to the theoretical models that they each base their views upon. For this reason, it is worth reviewing the dilemma represented by those who proclaim the need for a profound change of values and current lifestyles on the one hand, and, on the other hand, those whose overriding belief lies in the possibilities offered by technology and science.

If a bio-centric view is assumed, then a new ethic, new set of values, and new way of relating to nature are needed, including the search for technical solutions to change inappropriate (or nonsustainable) behavior (Robinson 2003, 844–45) being insufficient. We will now see that proposals have been advanced in line with a complete change of current consumer values (Oskamp 2000, 381). The majority of these indications are aimed at the attainment of sufficiency, in terms of a behavioral tendency to use less of what is technically, legally, and financially possible. In relation to behavioral changes in the social sciences (particularly social psychology), we also find indications of a more technical nature (Oskamp 2000; Princen 2003, 35).

The measures proposed previously need to be underpinned by a global intention. Thus, it would be highly dangerous (as well as dishonest) to propose austerity measures for those who have attained the lowest levels of development (and have, therefore, contributed least to the current deterioration), while those who live in developed countries do not consider a move toward new values to be so important (therefore contributing even further to the current levels of deterioration).

From an anthropocentric viewpoint, it is important to trust in human beings' skills and capacity to find technical responses that will enable solutions to be found for the problems that plague us and threaten the sustainability of the planet. In addition to the technical efforts involved in this, it is important to include the contribution of the behavioral sciences, which also entail the promotion of values that result in sustainability and ways of implementing values on a global scale (Robinson 2003, 841). The concepts of cooperation (equal representation, public participation, transparency, the sharing of information, consensus) and efficiency (a discussion of labor, economic scales, specialization, increases in efficiency or the management of governments in a more business-like fashion, intensification, and conservation) are representative of this line of thought. The most specific solutions advanced by this perspective are as follow:

- Implementation of eco taxes and new charges that will enable environmental costs to be passed on in the prices of different end products (Agyeman and Evans 2003, 39)
- Elimination of subsidies awarded to agriculture and energy
- Trade markets that are structured on more local lines
- Accessible housing
- Recycling and renewable energies
- Efficient transport

- Agricultural systems driven by the community
- Public participation
- More information
- More efficiency

Within the predominant spirit of this line of sustainable development, cooperation is the prevalent focus and implies negotiating, reaching agreement, implementing, monitoring, resolving disputes, and building confidence (Princen 2003, 33). In the same way as we saw in the analysis of the previous approach, we should also point out here that the need for participation and considered reflections, as well as the capacity to achieve collective solutions, is also important. This also implies the governance and design of a new kind of society, although not necessarily with the radicalism evident in the analysis of the previous approach.

Does Sustainable Development Necessitate a Change of the Economic and Social System? Reform versus Revolution

Although we shall limit our discussion here to pointing out certain doubts concerning sustainable development, we will, nevertheless, advance some key reflections. Both the different critiques of sustainable development as well as the different applications of it invite us to consider if what is actually being proposed through this process is the reform of our current economic, political, and social framework. Similarly, these approaches lead us to ask if we are heading from a break with this framework toward the implementation of a new one that would enable us to meet the challenges presented by sustainability.

If we understand sustainable development as a patch to ensure that a socioeconomic system on the brink of eruption continues to function (with the minimum amount of tweaking to enable it to continue on the same lines), then this would have no meaning at all, as it would fail to slow down the process of environmental, social, and economic deterioration. This would merely signify a delay in the ultimate decision-making processes, thereby serving to gain more time before the collapse, but not solving the problem of sustainability. In this way, we are faced with one sole possibility: a "paradigmatic shift," and, therefore, a more profound change of values than that required by a simple technological tweaking of the means of production and consumption. We are referring to an "auto-revolution" or, paradoxically, a radical reappraisal of the needs of the system itself and its will to change and evolve for its own sake, in order to avoid the catastrophic consequences facing us before we suffer from them in a way that would be fatal.

Presented in this way, this view enables some authors to see the conceptual complexity of sustainable development more in terms of a political act than as a scientific concept, a point that further hinders its practical significance. In most cases, it is quite clear to see what needs to be done, but problems arise when political and social obstacles hinder the implementation of solutions that have been identified.

In all cases, whether in the field of scientific or political debates, or even in terms of social and economic practices, the application of sustainable development should help to reveal the contradictions that currently exist between the different active agents who maintain different values and different interests (which are at times diverging and rarely coincide with the general interest). They must, nevertheless, reach agreements that enable them to face the challenge of a future based on interdependence and agreement (Petit 2003; Prades et al. 2004 and 2005).

To a certain extent, advancing further on this point implies the greater participation of individuals, more debate, and new forms of public consultation and debate, which will enable different points of view to be expressed and taken into account after they have been considered collectively.

The Need for Participation and Transparency

Sustainability is not merely a concept that takes in different fields of action (social, economic, and environmental), but also one that demands an end to the fragmentation of decision-making processes. This means that social and environmental concerns are integrated into decisions concerning economics and safety (Robinson 2003, 847). It is also a concept that includes both present generations and those of the future, and that must include different participatory agents in the process: governments, economic agents, and citizens (especially farmers, fishers, workers, women, children and young people, indigenous populations, the scientific and technological communities, and NGOs), as stipulated in Agenda 21.

To be immersed in a process of sustainable development implies the greatest possible level of participation. And this participation needs the appropriate transparency, deliberation methods, and decision-making processes to properly enable participation. This requires advances and significant transformations in terms of governance and political organization: a process of collective decision making that is based on experts' knowledge, but that is not determined by them. New forms of social education are also needed that will help to extend sustainable development to the different environmental, social, and political settings (Robinson 2003, 852–54).

In any event, we cannot ignore the fact that the key agents involved, and those that carry the loudest voice of the day, are capital and the means of production. It does not seem particularly likely that there can be any sustainable development without a real alternative to the needs of the agents that currently drive the economic and financial system, and without these agents collaborating fully in this development. With economic activity being fundamental for social prosperity, and with technology being one of the most vital instruments in this activity, we can begin to glimpse the importance that utilitarian tendencies and the instruments of efficiency and technological development will take on for those who advocate them.

Conclusion

For a long time it has seemed as though the socioeconomic system was omnipotent and did not depend upon anyone or anything. It seemed to be beyond good and evil and even beyond social and environmental realities, which were reduced to mere providers of resources to be exploited. We thought we could construct those environmental and social realities at will, in such a way that they would better serve the socioeconomic interests marked out by neoliberal capitalism. Now, without losing sight of the fact that the socioeconomic field influences and even shapes the other two areas (environmental and social), we find it disconcerting to note that the economic system is part of these two realities; it is not independent from them, but is rather an intrinsic part of them. In particular, we find that people and groups of people are not simply elements to be "used and discarded" within a blind market that includes them and uses them with the sole objective of obtaining the maximum economic profit from them (something that we have also discovered of late with respect to the environment).

It is, therefore, important to clarify what sort of relationship should exist between globalization and sustainable development (we have already seen that both phenomena are closely linked, even though this relationship can sometimes appear to be contradictory). According to our analysis, sustainable development must set the guidelines in this dialectic struggle. And this must be so due to the fundamental differences between both processes with regards to the objectives that each of them establishes and their resultant consequences. It is important to remember that sustainable development, by definition, requires the participation of all of the social and individual agents involved in the difficult situations of our present day and that this, in and of itself, means that the process is of greater importance than globalization.

The latter is governed by mechanisms that are blind to the consequences that it produces as well as opaque in terms of decision making and participation; exactly the opposite of sustainable development. Globalization seeks to impose the economic interests that it is driven by on other nondominant economic interests, as well as on different social and environmental situations that, in this way, remain subordinated to the neoliberal capitalist economy. Sustainable development, by contrast, implies the pursuit of a socioeconomic system that is in harmony with the social and cultural needs of the people who are part of it. It also implies a scrupulous respect for the environment as a source of resources upon which any such sustainable socioeconomic system rests and depends.

At this point, and in order to conclude, we would like to highlight the key points that emerge from the analysis and study carried out here. These points indicate the directions to be taken in the future—if this is not already occurring or has not always occurred in this way—with regard to the questions covered in this study.

What we find is that while advances have been visibly made with respect to the environment and in terms of the efficiency of sustainable development, much remains to be done in terms of the social side of this equation. Apparently the point needs to be highlighted that a society that forgets this aspect cannot be sustainable. In this sense, we believe that much more effort and considerable reflection is needed to help us see the need (even economically speaking) to attain social sustainability (with all that that implies) with the same level of vigor and urgency as that which is being applied in the other two more environmental and economic parameters of sustainable development.

Equally, in the future we are going to see a greater debate between preservationist and conservationist tendencies, between those who advocate maintaining and improving the system and those who call for the need for a radical change to the socioeconomic system. This debate will not only be theoretical, but will also extend beyond the field of ideas into the realm of the practical and undoubtedly be seasoned with tensions and fierce discussion. In any event, we will see that both approaches can be equally understood and merged together, with both approaches mutually complementing each other.

The third key point in this question will continue to be participation and the political and governmental changes that are necessary to maintain a truly sustainable planet. It remains to be seen if advances in these two crucial issues may help us to clarify this question and to provide it with a satisfactory solution that would enable the entire human species to reach the new utopia proposed. We are referring here to the new Promised Earth, the pursuit of which we must follow if we wish to survive, and some of the keys to which can be found in sustainable development.

References

Agyeman, Julian, and Tom Evans. 2003. "Toward Just Sustainability in Urban Communities: Building Equity Rights with Sustainable Solutions." *Annals of the American Academy of Political and Social Science* 590, no. 1 (November): 35–53.

Albert, Michel. 1993. *Capitalism against Capitalism*. Translated by Paul Haviland. Foreword by Lord Donoghue. London: Whurr.

Blanchflower, David G. 2005. "Happiness and the Human Development Index: The Paradox of Australia." *Australian Economic Review* 39, no. 2: 176–84.

Blanchflower, David G., and Andrew J. Oswald. 2004. "Wellbeing over Time in Britain and the USA." *Journal of Public Economics* 88, nos. 7–8: 1359–86.

Campbell, Catherine, and Sandra Jovchelovitch. 2000. "Health, Community and Development: Towards a Social Psychology of Participation." *Journal of Community and Applied Social Psychology* 10, no. 4: 255–70.

City of Winnipeg. 1995. *A Community Indicators System for Winnipeg*.

Comisión Económica para América Latina y el Caribe(CEPAL). 2002. *Globalización y Desarrollo*. Brasilia: Naciones Unidas.

Cuthill, Michael. 2002. "Exploratory Research: Citizen Participation Local Government and Sustainable Development in Australia." *Sustainable Development* 10, no. 2: 79–89.

Doelle, Meinhard, and A. John Sinclair. 2006. "Time for a New Approach to Public Participation in EA: Promoting Cooperation and Consensus for Sustainability." *Environmental Impact Assessment Review* 26, no. 2: 185–205.

Fernando, Jude L. 2003. "Preface: The Power of Unsustainable Development; What Is to Be Done?" *The Annals of the American Academy of Political and Social Science* 590, no. 1 (November): 6–34.

Harrison, Carolyn, Richard J.C. Munton, and Kevin Collins. 2004. "Experimental Discursive Spaces: Policy Processes, Public Participation and the Greater London Authority." *Urban Studies* 41, no. 4: 903–17.

Keeling, David J. 2004. "Latin American Development and the Globalization Imperative: New Directions, Familiar Crises." *Journal of Latin American Geography* 3, no. 1: 1–21.

Keyes, Corey Lee M. 1998. "Social Wellbeing." *Social Psychology Quarterly* 61, no. 2: 121–40.

Klein, Naomi. 2000. *No Logo-Taking Aim at the Brand Bullies*. Toronto: Vintage.

Lohmann, Larry. 1990. "Editorial: Whose Common Future?" *Ecologist* 20, no. 3 (May/June): 82–84.

Lucas, S.E. Manuela. 2004. "How Can We Make Globalization Work for Sustainable Development?" *Afrique PALOP: Développement et mondialisation*. Seminar, Université Libre de Bruxelles, March 1–2. At www.amigos-de-mocambique.org/sem/PALOP-Lucas.pdf.

McAllister, Fiona. 2005. *Wellbeing Concepts and Challenges*. Sustainable Development Research Network Discussion Paper. December. At www.sd-research.org.uk/wp-content/uploads/sdrnwellbeingpaper-final_000.pdf.

Morton, Brian. 2000. "Sustainable Development." *Marine Pollution Bulletin* 40, no. 7: 563–64.

Norgaard, Richard B. 1988. "Sustainable Development: A Co-evolutionary View." Editorial. *Futures* 20, no. 6 (December): 606–20.

———. 1994. *Development Betrayed: The End of Progress and a Coevolutionary Revisioning of the Future*. New York: Routledge.

Oskamp, Stuart. 2000. "Psychological Contributions to Achieving an Ecologically Sustainable Future for Humanity." *Journal of Social Issues* 56, no. 3: 373–90.

Petit, Michel. 2003. "Développement durable à l'échelle de la planète et gestion des ressources en eau et en sols." *C.R. Geoscience* 335: 643–56.

Prades, José A., R. Belzile, M. Glaus, R. Hausler, Jean-Pierre Revéret, and Juan Antonio Rubio-Ardanz. 2004. "Norms and Values in Contemporary Society: An Application to the Field of Sustainable Transportation." In *Human Perspectives in the Internet Society: Culture, Psychology and Gender*, edited by Konrad Morgan, J. Sanchez, Carlos A. Brebbia, and Alexander Voiskounsky. Boston: WIT Press.

Prades, José A., Martin Mujica, Jean-Pierre Revéret, and Juan Antonio Rubio-Ardanz. 2005. "Vers une stratégie de transport durable fondée sur le développement de l'innovation technologique." *Esprit critique* 7, no. 1 (Winter): 52–65.

Princen, Thomas. 2003. "Principles for Sustainability: From Cooperation and Efficiency to Sufficiency." *Global Environmental Politics* 3, no. 1: 33–50.

Robinson, John. 2003. "Future Subjunctive: Backcasting as Social Learning." *Futures* 35, no. 8: 839–56.

Rodrik, Dani. 2001. *The Global Governance of Trade as if Development Really Mattered*. United Nations Development Programme Background Paper. New York: United Nations Development Programme. At www.wcfia.harvard.edu/sites/default/files/529__Rodrik5.pdf.

World Commission on Environment and Development. 1987. *Our Common Future*. Oxford: Oxford University Press.

Index

List of Contributors

For full biographical information about the contributors, links to their projects, and more, visit www.basque.unr.edu/currentresearch/contributors.

Ibone Amezaga

Juan Antonio Rubio-Ardanaz
University of Extremadura

Xabier Arana
Environment Department, Regional Government of Biscay

Arantza Aranburu

M. Arriolabengoa

Ignacio Ayestarán

Roberto Bermejo

Izaskun Casado-Arzuaga

Alejandro Cearreta

Laura Damas-Mollá

Agustín Ezcurra

Xiao Fang
Lu Xun Center for Chinese Studies, Bilbao

Patxi García-Garmilla

Eneko Garmendia
Postdoctoral Researcher at Basque Centre for Climate Change (BC3)

David Hoyos

Gabriel Ibarra-Berastegi

Eneko Iriarte
University of Cantabria

Peru Iridoy

Maria Jesús Irabien

Eduardo Leorri
East Carolina University

Iosu Madariaga
Environment Department, Regional Government of Biscay

Teresa Nuño Angós

Miren Onaindia

Igone Palacios

Lorena Peña

Pío Pérez Aldasoro

Gloria Rodríguez-Loinaz

Eduardo Rubio Ardanaz

Jon Sáenz

Eguzki Urteaga

Araitz Uskola Ibarluzea

Iñaki Yusta

Made in the USA
Monee, IL
04 November 2026

67387542R00125